D0482347

WITHDRAWN

The Puzzle Instinct

The Puzzle Instinct

The Meaning of Puzzles in Human Life

Marcel Danesi

The Puzzle Instinct

The Meaning of Puzzles in Human Life

Marcel Danesi

INDIANA
University Press
Bloomington & Indianapolis

This book is a publication of

Indiana University Press
601 North Morton Street
Bloomington, Indiana 47404-3797 USA

http://iupress.indiana.edu

Telephone orders 800-842-6796
Fax orders   812-855-7931
Orders by e-mail iuporder@indiana.edu

© 2002 by Indiana University Press
All rights reserved

No part of this book may be reproduced or utilized in any
form or by any means, electronic or mechanical, including
photocopying and recording, or by any information storage
and retrieval system, without permission in writing from the
publisher. The Association of American University Presses'
Resolution on Permissions constitutes the only exception
to this prohibition.

The paper used in this publication meets the minimum
requirements of American National Standard for Information
Sciences—Permanence of Paper for Printed Library
Materials, ANSI Z39.48-1984.

MANUFACTURED IN THE UNITED STATES OF AMERICA

Library of Congress Cataloging-in-Publication Data

Danesi, Marcel, date
  The puzzle instinct : the meaning of puzzles in human life / Marcel
Danesi.
     p. cm.
Includes bibliographical references (p.   ) and index.
  ISBN 0-253-34094-2 (cloth : alk. paper)
  1. Puzzles—Psychological aspects.   2. Puzzles—Social aspects.   3.
Games—Psychological aspects.   4. Games—Social aspects.   I. Title:
Meaning of puzzles in human life.   II. Title.
  GV1493 .D337 2002
  793.73—dc21
                                      2001007492

1   2   3   4   5   07   06   05   04   03   02

To Sarah and Alexander

Your sparkling eyes reveal understanding and wisdom well beyond your years. May you always be able to solve the puzzles that life will present to you with insight and imagination.

Contents

5 Puzzling Numbers   143
Magic Squares, Cryptarithms,
and Other Mathematical Recreations

6 Puzzling Games   178
Chess, Checkers, and Other Games

7 The Puzzle of Life   208

PREFACE

Puzzles, like virtue, are their own reward.
—Henry E. Dudeney (1847–1930)

Why does everyone love a good puzzle, as the expression goes? Why do puzzle magazines, brain-challenging puzzle sections in newspapers, and game tournaments appeal to hordes of ordinary people from every walk of life? Is this widespread penchant for puzzles no more than a passing fad foisted by our market culture on a gullible, bored populace spoiled by consumerism? Hardly. Puzzles have been around since the dawn of history. One of the oldest mathematics textbooks (the Egyptian *Rhind Papyrus,* which dates back to 1650 B.C.) turns out to be essentially a collection of mathematical brainteasers. The biblical kings Solomon and Hiram were renowned for organizing riddle contests. Benjamin Franklin, the American inventor, and Lewis Carroll, the English writer known for his two great children's novels, *Alice's Adventures in Wonderland* and *Through the Looking-Glass,* contrived ingenious puzzles that intrigue people to this day.

Why have people from time immemorial been so fascinated by seemingly trivial posers, which nonetheless require substantial time and mental effort to solve, for no apparent reward other than the simple satisfaction of solving them? Is there a *puzzle instinct* in the human species, developed and refined by the forces of natural selection for some survival function? Or is this instinctual love of puzzles the product of some metaphysical force buried deep within the psyche, impelling people to behave in ways that defy rational explanation?

My purpose in writing this book has been to tackle such questions head on. It has not been conceived as a historiography of puzzles, although the histories of some puzzle genres will be discussed when they are relevant. Rather, I have written it to convey a sense of the intriguing story that puzzles tell about human life. Will Shortz, the current *New York Times* crossword puzzle editor (and former editor of *Games Magazine*), has called for a systematic study of the relation between puzzles and culture under the rubric of *enigmatology.* This book is intended to fall under Shortz's rubric, since it aims to shed some light on the meaning of puzzles in everyday human life by searching for, and unraveling, the sources of the puzzle genres that have withstood the test of time.

My own fascination with puzzles started in childhood. I would often become obsessed with finding the answer to a riddle or discovering the correct path through a maze that I found in a magazine. As an adolescent in the early 1960s, I read Martin Gardner's monthly puzzle column in *Scientific American* religiously, following up in my local library the enticing historical leads that Gardner would provide and finding, to my constant surprise, that some classic puzzles originated in the context of an ancient magical or occult tradition. Writing this book has itself been akin to solving a puzzle, for I had to bring together many textual sources, tidbits of information scattered here and there, and then attempt to flesh out some general pattern within the sources and the information. I am not sure whether I have found any such pattern to or meaning in puzzles. Maybe there is none. All I can say is that the search for one has been revealing and pleasurable. I hope I can impart to the reader some of the pleasure that investigating puzzles has accorded me.

I would like to thank all those who have given me feedback, advice, and encouragement during the writing of this book. First, I am extremely grateful to John Gallman and Peter-John Leone of Indiana University Press for their kindly interest in my work; without their support, this book would never have been contemplated in the first place. I am deeply indebted to both Dr. Raymond Smullyan (one of the greatest puzzlists of all time and a primary target of interest of this book), Keith Devlin (a leading mathematical thinker, writer, and educator), and Will Shortz (a worldwide authority on puzzles and their history) for their comments, constructive critiques, and extremely helpful suggestions on the manuscript. These have allowed me to improve it considerably. I am especially grateful to Shoshanna Green, whose thoughtful and thorough editing of the manuscript has truly transformed it into a highly readable, coherent work. Needless to say, any infelicities that this book may still contain are my sole responsibility. I also wish to express my gratitude to Victoria College of the University of Toronto for a research grant that allowed me to compile the appropriate background textual sources. Finally, I must thank my good friend and colleague Professor Ed Barbeau of the Department of Mathematics (University of Toronto) for all the insights on the nature of puzzles and problem-solving he has given me over the years.

Marcel Danesi
University of Toronto, 2001

The Puzzle Instinct

1 Why Puzzles?

Man is a puzzle-solving animal.
—Ronald A. Knox (1888–1957)

The great American author Henry David Thoreau (1817–1862) once remarked that, for some enigmatic reason, human beings require that things be mysterious. Supporting Thoreau's perceptive observation is the worldwide popularity of suspense thrillers, horror stories, and detective novels. But the penchant for mystery is not merely modern. Literary historians trace the origin of the horror narrative genre to *The Castle of Otranto,* written by Horace Walpole (1717–1797) in 1764, and of the detective story to "The Murders in the Rue Morgue," written by Edgar Allan Poe (1809–1849) in 1841. And long before Walpole and Poe, people satisfied their appetite for mystery in many other ways. The mystery plays of medieval times kept spectators in suspense by reenacting before their eyes the miracles performed by the saints. In the ancient world, the dramas of Aeschylus (525–456 B.C.), Sophocles (c. 496–406 B.C.), and Euripides (c. 480–406 B.C.) held Greek audiences spellbound, as they watched with apprehension the fateful actions of legendary heroes and heroines on stage. Scholars now believe that the drama genre itself may have developed from the rituals performed by the secret mystery cults of ancient Greece (Mishlove 1993: 40), in which Pythagoras (c. 582–c. 500 B.C.), Plato (c. 428–c. 347 B.C.), and other great thinkers are said to have taken part. The central purpose of those rituals appears to have been to pose questions about the mystery of existence (Hall 1973).

As far as can be determined, no other animal species displays a comparable need for mystery and suspense. So, it may be asked, what purpose does it serve in human life? Plato eventually came to believe that this peculiar need served no purpose whatsoever, arising simply out of superstitious traditions which, he asserted, put obstacles in the path of true science. After observing the method of philosophical inquiry used by the Athenian philosopher and teacher Socrates (c. 470–399 B.C.), whom he greatly admired, Plato proclaimed *dialectic reasoning* to be the only useful method of gaining knowledge, defining it as the Socratic practice of examining ideas logically by means of a sequence of questions and answers. But Plato apparently ignored the fact that the central practices of the "superstitious traditions" he denounced were themselves fundamentally dialectic in nature, since their aim was to probe the mystery of existence by posing questions about it.

Puzzles emerged at about the same time as the mystery cults—right at the dawn of human history. As will become obvious in the course of this book, I do not believe this to be a mere coincidence. Puzzles and mysteries are intrinsically intertwined in human life. Suffice it to say at the outset that they appeal to people for the very same reason—they generate a feeling of suspense that calls out for relief. The word *catharsis* was used by Aristotle (384–322 B.C.) to describe the sense of emotional relief that results from watching a tragic drama on stage. Unraveling the solution to a mystery story or to a puzzle seems to produce a kind of "mental catharsis," since people typically feel a sense of relief from suspense when they find the answer to the mystery or puzzle. For some truly mysterious reason, human beings seem to need this kind of catharsis on a regular basis—as the popularity of mysteries and puzzles across the world and across time attests.

Why is this so? To the best of my knowledge, this question has rarely been formulated, let alone investigated. The goal of this book is, in fact, to investigate this and the many other fascinating questions that the existence of puzzles in human life raises. One of the greatest puzzlists of all time, Lewis Carroll (the pseudonym of Charles Dodgson, 1832–1898), saw puzzles as structures born of the imagination. In his two masterpieces of children's literature, *Alice's Adventures in Wonderland* and *Through the Looking-Glass and What Alice Found There,* he portrayed the imagination as a region called Wonderland, inhabited by personified puzzle concepts such as Humpty Dumpty, Tweedledee and Tweedledum, the Cheshire cat, and others. In tribute to Carroll, the modern-day puzzlist Raymond Smullyan coined the term "puzzle-land" in his entertaining and intriguing 1982 book *Alice in Puzzle-Land* to describe the Carrollian realm of the imagination, where the puzzle instinct resides. Smullyan's term is an appropriate metaphor for the domain which this book intends to look into.

Puzzle-Making through the Ages

When asked why puzzles are so popular, most people answer that, like any other hobby, they provide a form of constructive entertainment during leisure hours. But their unparalleled popularity through the ages speaks of something much more profound than just recreation. Some of the oldest and most widely read books of history are collections of puzzles and games. The *Book of Games,* to mention but one, was one of the best-selling books of the entire medieval period (Mohr 1993: 11). Commissioned by King Alfonso X of Castile and Léon (c. 1226–1284), and containing clear descriptions of how to play chess, checkers, and various card and board games, it appealed to aristocrats and common folk alike. Today, puzzles are among the most profitable commercial products manufactured. Over two hundred million Rubik's Cubes were sold in the early 1980s alone. Every year at Christmas time toy stores cannot keep up with the demand for jigsaw puzzles, game boards, Lego sets, and the like. There appear to be more specialized periodicals devoted to puzzles than to any other single area of human interest and concern. And few other hobbies attract as many devotees as do puzzles. Associations such as the National Puzzlers' League, founded in the United States at the end of the nineteenth century, are springing up all over the world. So intense is some people's desire for puzzles that psychologists have even identified syndromes such as "puzzle fixation" and "puzzle depression," associated with a fierce, irrational craving for puzzles. Clearly, there is much more to puzzles than amusement.

As legend has it, the first "intelligence test" ever devised was a puzzle. It has come down to us as the Riddle of the Sphinx (Grimal 1963: 324). In ancient mythology, the Sphinx was a monster with the head and breasts of a woman, the body of a lion, and the wings of a bird. Lying crouched on a rock, the Sphinx menacingly accosted all who dared enter the city of Thebes by posing a riddle:

What is it that has four feet in the morning, two at noon, and three at twilight?

The beast mercilessly killed on the spot anyone who failed to answer the riddle correctly. On the other hand, it had vowed to destroy itself if someone ever came up with the correct answer. When the hero Oedipus— who on his way to Thebes had killed a mysterious stranger—solved the riddle by answering, "Man, who crawls on four limbs as a baby [in the morning of life], walks upright on two as an adult [at the noon hour of life], and gets around with the aid of a stick in old age [at the twilight of life]," the Sphinx jumped from its perch to a rock outside the city, becoming a lifeless statue. For ridding them of the terrible monster, the Thebans made Oedipus the successor of their king, Laius, who had recently been murdered, and Oedipus married Laius's widow Jocasta. But the consequences of his ingenuity were disastrous for both Oedipus and Thebes. Several years later, a plague struck the city. An oracle announced that the scourge would come to an end only after Laius's murderer had been driven from Thebes. Oedipus investigated the murder and soon realized that Laius was the man he had killed on the road. To his horror, he also discovered that Laius was his father and Jocasta his mother. Grief-stricken and desperate, Oedipus blinded himself. Similarly woeful and disconsolate, Jocasta hanged herself. Oedipus was banished from Thebes. He died in a state of unendurable woe at Colonus, near Athens.

Throughout the ancient world, puzzles were associated with portentous challenges and events. This association was given physical expression in the form of buildings known as *labyrinths,* which were, in effect, architectural intelligence tests. Finding one's way through their intricate, intertwining passages was considered not only a test of astuteness, but also a way of metaphorically finding the path to enlightenment and true knowledge. One of the most celebrated labyrinths was the funeral temple constructed by Amenemhet III in Egypt, which contained three thousand chambers. Equally famous was the prison built on the island of Crete, which may, however, have existed only in myth (chapter 3). According to legend, Androgeus, the son of King Minos of Crete, was killed, apparently by Athenians. To exact his revenge, Minos hired the craftsman Daedalus to build a dungeon in the form of a labyrinth that would house the Minotaur—a creature who was half human and half bull—at its center. Fourteen young Athenians, seven men and seven women, were to be sent into the prison every year, so that the Minotaur could destroy them. Athens was, however, saved by Theseus, son of King

Aegeus, with the help of his clever lover Ariadne—who was, ironically, Minos's daughter. The brave hero killed the Minotaur and then found his way out of the twisting passages by simply following the thread given to him by Ariadne, which he had unwound behind himself on his way in. Archaeologists have discovered a palace located in the Cretan city of Knossos that may have been the site of the mythical labyrinth, because it has many passageways like those described in the legendary account of the building.

While the mystery surrounding labyrinths may have faded, they continue to be built as tests of wits, both recreational and serious. Maze structures with confusing, intertwining networks of passages are commonly found in amusement parks. These hold people in suspense as they try to figure out how to find their way through them. Psychologists, on the other hand, frequently use mazes more seriously to test problem-solving skills in animals and humans alike. And toy mazes are among the most popular types of games given to children today, mainly because they are thought to sharpen children's logical skills, while providing recreation at the same time.

According to some scholars, puzzles may even be older than recorded history. In a fascinating book titled *Ancient Puzzles,* Dominic Olivastro (1993: 5–11) speculates that around eleven thousand years ago a tribe living near Lake Edward in modern-day Zaire, who were the ancestors of the Ishango, invented what appears to be the first mathematical game of humanity, consisting of two "dice": bones on which notches represent numbers. If Olivastro is right, then the puzzle instinct is ancient indeed. This would explain, perhaps, why one of the earliest surviving manuscripts of human civilization is, as a matter of fact, a collection of mathematical puzzles. Eighteen and a half feet long and thirteen inches wide, the manuscript is referred to either as the *Ahmes Papyrus,* after the Egyptian scribe A'h-mose, or Ahmes, who copied it, or the *Rhind Papyrus,* after a certain Scottish lawyer and antiquarian, A. Henry Rhind (1833–1863), who purchased it in 1858 while vacationing in Egypt. The papyrus had been found a few years earlier in the ruins of a small building in Thebes in Upper Egypt.

As he tells us in his introductory remarks, Ahmes' manuscript is a copy of an older anonymous work (Chase 1979: 27):

> This book was copied in the year 33, in the fourth month of the inundation season, under the majesty of the king of Upper and Lower Egypt, "A-user-Re," endowed with life, in likeness to writings

made in the time of the king of Upper and Lower Egypt, "Ne-ma'et-Re." It is the scribe A'h-mose who copies this writing.

The king A-user-Re has been identified as a member of the Hyskos dynasty, which was in power around 1650 B.C. The king Ne-ma'et-Re was Amenemhet III, who reigned from 1849 to 1801 B.C. Thus, the original manuscript was written nearly four thousand years ago, during the same period in which another famous record of Egyptian mathematics, the *Moscow Papyrus* (named after its current location), was written. In addition to eighty-four challenging mathematical problems, the Rhind Papyrus contains tables for the calculation of areas and the conversion of fractions, elementary sequences, linear equations, and extensive information about measurement. The earliest known symbols for addition, subtraction, and equality are also found in this truly remarkable work. The Moscow Papyrus, on the other hand, contains only twenty-five problems, which are less interesting than those found in the Rhind Papyrus (Gillings 1972: 246–247).

The Rhind Papyrus was translated into German in 1877 by a mathematician named August Eisenlohr (Blatner 1997: 10), and into English in 1923 by a British scholar named Thomas Eric Peet. The first extensive edition of the work was completed in 1929 by Arnold Buffum Chase (1845–1932)—an edition that made the Rhind Papyrus accessible for the first time to the general public (Chase 1979). The original papyrus is now preserved in the permanent collection of the British Museum, which came into possession of it after Rhind's untimely death at the age of thirty. The papyrus was missing certain fragments, but by a stroke of incredible fortune, the fragments were found a little later in the possession of the New-York Historical Society.

The work starts off with the following little poem:

Accurate reckoning,
the entrance into the knowledge of all existing things and all
*   obscure secrets.*

This poetic preface has suggested to some that the papyrus was intended as a practical manual—with eighty-four problems in arithmetic, geometry, and algebra, together with their solutions—for Egyptian youths struggling with mathematics. Perhaps it was meant to impress upon them, by pedagogical illustration, that the exercise of accurate reckoning was the key to successful problem-solving (Olivastro 1993: 31–64; Gillings 1961, 1962, 1972). But the counterargument can be made that the work simply contains too many original ideas to have been designed exclusively as a pedagogical tool. Solving Problem 56, for instance, leads to an insight that was original for the era in which the papyrus was written—namely, that the height of a pyramid can be related to the size and slope of each of its triangular walls. Even the papyrus's title—*Directions for Attaining Knowledge of All Dark Things*—speaks more of something shrouded in mystery than of something purely educational. Mystery, wisdom, and puzzle-solving were intrinsically intertwined in the ancient world. Many of the first books of wisdom turn out, upon closer scrutiny, to be elaborate puzzle creations. The *I Ching*, for instance, which was written during the Shang dynasty of ancient China (c. 1766–c. 1122 B.C.) and which has traditionally been used for divination, is organized around sixty-four symbolic hexagrams that are miniature puzzles explained by texts that can be characterized as "cryptic poems." The *I Ching* uses two basic lines—a *yin* (broken line) and a *yang* (unbroken line)—to draw the symbols. The lines are converted into numbers and then into symbolic answers to spiritual questions.

Above all else, the Rhind Papyrus is an archaeological treasure, so to speak, for a systematic study of puzzles. It contains concepts that have surfaced in other areas of the world and in languages that are unrelated to the language used by Ahmes. Here, for instance, is the papyrus's Problem 79, which it presents in the form of an inventory without an appurtenant question:

houses	7
cats	49
mice	343
sheaves of wheat	2,401
hekats of grain	16,807
estate	19,607

Why the puzzle was presented in this way is not known. Perhaps Ahmes simply forgot to write out the question while copying the original manuscript; or he may have thought that the question was too obvious to be written down. Whatever the case, one thing is clear—the creator of this puzzle had a specific numerical concept in mind, since the first five numbers are successive powers of 7: i.e., $7 = 7^1$, $49 = 7^2$, $343 = 7^3$, $2{,}401 = 7^4$, and $16{,}807 = 7^5$; while the last figure, 19,607, is the sum of these numbers: $7 + 49 + 343 + 2{,}401 + 16{,}807 = 19{,}607$. What we seem to have here is a solution without a puzzle statement, with the first five numbers representing calculations and the final figure their sum.

The very same puzzle appears in other parts of the world at later times. It surfaces, for instance, in *The Book of the Abacus,* written by the medieval Italian mathematician Leonardo Fibonacci (c. 1170–c. 1240). Fibonacci added, however, another power of 7, namely 7^6, to his version of the puzzle:

> *Seven old women are on the road to Rome. Each woman has seven mules, each mule carries seven sacks, each sack contains seven loaves, to slice each loaf there are seven knives, and for each knife there are seven sheaths to hold it. How many are there altogether: women, mules, sacks, loaves, knives, sheaths?*

The solution to Fibonacci's puzzle can be shown simply as follows:

number of women	=	7^1	=	7
number of mules	=	7^2	=	49
number of sacks	=	7^3	=	343
number of loaves	=	7^4	=	2,401
number of knives	=	7^5	=	16,807
number of sheaths	=	7^6	=	117,649
	total:			137,256

Fibonacci could not possibly have known about Ahmes' puzzle, because the existence of the papyrus was not known at the time. But the similarity between the two is unmistakable. With Fibonacci's version, it is easy to reconstruct an appropriate wording for Problem 79. One possibility is as follows:

> *A man owns an estate with 7 houses on it. In each house he keeps 7 cats. Each cat chases 7 mice. Each mouse nests in 7 sheaves of wheat. Each sheaf of wheat produces 7 hekats of grain. How many houses, cats, mice, sheaves, and hekats are there on the estate, all told?*

In eighteenth-century England this same puzzle appeared as a popular nursery rhyme:

> *As I was going to St. Ives*
> *I met a man with seven wives.*
> *Each wife had seven sacks,*
> *Each sack had seven cats,*
> *Each cat had seven kits.*
> *Kits, cats, sacks, wives,*
> *How many were going to St. Ives?*

That version, however, contained a clever trap. The sly anonymous puzzlist asked how many kits, cats, sacks, and wives were *going to* St. Ives, not *coming from* it. Only one person was going to St. Ives—the narrator of the poem. All the others were, of course, making their way out of the city.

Why has Ahmes' puzzle concept cropped up in various places and at different times across the world? One possibility is that the number 7 has a strange appeal to people, as witnessed by the fact that it is commonly imbued with mystical connotations. There are seven gods of good fortune in Japanese lore; seven chieftains in Greek mythology who undertook an ill-fated expedition against the city of Thebes; seven deadly sins, according to Catholic medieval theologians; seven cosmic truths, according to the Plains Indians of North America; and the list could go on and on (Chevalier and Gheerbrant 1994: 859–868). The ubiquity of this puzzle can perhaps be explained as resulting from the unconscious symbolic appeal that the number 7 seems universally to possess. On the other hand, Gillings (1972: 168) offers a more mundane reason for Ahmes' use of the number 7: "The number 7 often presents itself in Egyptian multiplication because, by regular doubling, the first three multipliers are *always* 1, 2, 4, which add to 7." But, as Maor (1998: 13) remarks, this explanation is unconvincing because "it would equally apply to 3 (= 1 + 2), to 15 (= 1 + 2 + 4 + 8), and in fact to all integers of the form $2^n - 1$." Maor's own follow-up explanation is also unconvincing: "7 was chosen because a larger number would have made the calculation too long, while a smaller one would not have illustrated the rapid growth of the progression." Given the Egyptian numeration system of Ahmes' times, the fifth term of the progression in the puzzle, $16{,}807 = 7^5$, would hardly have been perceived as a convenient number to work with for either calculation or illustrative purposes.

The creation of puzzles with numerological symbolism in mind was not unusual in the ancient world. Such symbolism also characterized the emerging science of mathematics. From all accounts, the demonstration that the area of the square drawn on the hypotenuse of a right-angled triangle is equal to the sum of the areas of the squares drawn on the other two sides—or, in symbolic form, $c^2 = a^2 + b^2$, in which c is the length of the hypotenuse and a and b the lengths of the other two sides— was felt to be an event of great mystical significance by its demonstrator, Pythagoras [see figure 1.1].

Legend has it that Pythagoras believed the divinities had allowed him, a mere mortal, to catch a unique glimpse into the *raison d'être* of

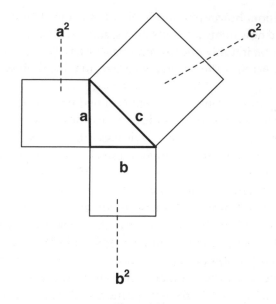

Figure 1.1

one of the numerical laws governing the cosmos—the word "cosmos" meant "good array" in Greek and, by metaphorical extension, "universe." Pythagoras is purported to have offered a sacrifice to the gods for granting him "entrance into the knowledge of an obscure secret," as Ahmes might have characterized his accomplishment. He also apparently demanded that his followers take an oath, on pain of death, not to reveal the proof of the theorem—known ever since as the Pythagorean Theorem. But even before Pythagoras, the $c^2 = a^2 + b^2$ relation was known in other lands. One of its earliest demonstrations appeared around six hundred years before Pythagoras, and in a totally different part of the ancient world—in *The Arithmetic Classic of the Gnomon and the Circular Paths of Heaven,* a Chinese treatise that is dated to around 1100 B.C. (Li Yan and Du Shiran 1987). The Pythagorean Theorem shows up, in fact, in manuscripts throughout the ancient world in regions where it is unlikely that the work of Pythagoras would have been known. It turns up, for example, in *Nine Chapters on the Mathematical Art,* a Chinese collection of puzzles compiled sometime in the third century B.C. (Swetz and Kao 1977).

At this early point in our journey through Puzzleland, a caveat is in order. It would be misleading to convey the impression that the ancient mathematicians and puzzle-makers had only noble objectives in mind when they went about their work. As it turns out, their motives were often much less commendable, and certainly much more mundane. The great Sicilian mathematician and scientist Archimedes (287–212 B.C.), for instance, invented arithmetical puzzles not only to probe truths about the world but, on occasion, to befuddle his foes. Legend has it that he contrived his famous Cattle Problem (chapter 5) as a way of taking revenge on one of his adversaries, whom he obviously was trying to dumbfound with his computational dexterity. Nevertheless, as history records, even that impishly devised puzzle has attracted, at some time or other, the attention of virtually every great mathematician, and has been the source of many mathematical insights.

The list of ancient scholars who devised puzzles is extensive. Heron of Alexandria (c. A.D. 20–c. 62), for example, who was known for mechanical inventions such as a rotary steam engine and a primitive surveying instrument, also contrived ingenious puzzles as frameworks for investigating square and cube roots. In his famous treatise *Arithmetica*, the great Alexandrian mathematician Diophantus (c. A.D. 200–c. 284) constructed puzzles specifically to illustrate methods of solving algebraic equations. One type of puzzle that he created to do so has come to be known as *Diophantine*. It will be discussed below.

Some of the most popular works of the medieval era were puzzle collections. Early in the period the *Greek Anthology*, a compilation of literary verses, epigrams, riddles, enigmas, and mathematical puzzles, became known far and wide. Some credit the poet Metrodorus (c. A.D. 500–?) with writing it, although the author is not known for certain. The *Anthology* is interesting historically because it contains many of the same puzzle concepts found in the Rhind Papyrus and in other ancient works. Here is one of its problems (Wells 1992: 23):

> *I desire my two sons to receive the thousand staters of which I am possessed, but let the fifth part of the legitimate one's share exceed by ten the fourth part of what falls to the illegitimate one.*

Using modern algebraic notation, this puzzle can be solved easily today (see solution 1.1 in the solution section at the back of this book), but one cannot help but wonder how people in the sixth century A.D. went about solving it. Problems of this type are now included routinely in introductory mathematics textbooks, having become standard ways of teaching how to solve "story problems" in algebra. But in the early medieval era, when modern-day algebraic techniques were unknown, their solutions required a much more extensive use of ingenuity—the hallmark of true puzzle-solving. Today, Metrodorus's problem would hardly be classified as a genuine puzzle, but rather as a simple algebraic problem.

One of the first "puzzle addicts" of history was none other than Charlemagne (A.D. 742–814), the founder of the Holy Roman Empire, who became so obsessed with puzzles that he hired an expert full time to create puzzles for him. The person he selected for the job was the well-known English scholar and ecclesiastic Alcuin (A.D. 735–804). Because of his position in Charlemagne's court, Alcuin succeeded in establishing an effective educational program among the Franks, thereby exercising lasting influence upon the intellectual life of the Western world. The ingenious Alcuin put together fifty-six of the puzzles he invented for Charlemagne into an instructional manual, titled *Problems to Sharpen the Young,* for training medieval youth in mathematics and accurate reckoning. Below is one of the clever puzzles that the *Problems* contains:

> *When 100 bushels of grain are distributed among 100 persons so that each man receives 3 bushels, each woman 2 bushels, and each child ½ bushel, how many men, women, and children are there?*

This is an example of a Diophantine puzzle, defined as an algebraic problem in which there are more unknowns (here, three) than there are equations (two). The number of solutions is thus theoretically infinite, but its unique practical solution can be wrested out of the two equations on the basis of the given facts (see solution 1.2). Incidentally, the same puzzle shows up in different guises in other parts of the world at other times. For instance, it appears much earlier in the third-century A.D. *Bhakshali Manuscript,* discovered in northwest India in 1881.

Although the ancient Egyptians, Babylonians, Chinese, and peoples of India knew the notion of *variable,* it was Diophantus who spread the practice of using symbols for unknown numbers in the ancient world. This is why he is often called "the father of algebra." His ideas were developed throughout the medieval ages primarily by Arab scholars such as the mathematician Al-Khwarizmi (c. 780–c. 850), a teacher in Baghdad, and, later, the Persian astronomer and poet Omar Khayyam (c. 1048–1123). The English word "algebra," in fact, comes from the word *al-jabr,* meaning "restoration" or "reunion," in the title of Al-Khwarizmi's work, *Calculation by Restoration and Reduction* (in Arabic, *Hisab al-jabr w'al-muqabala*). And the term "algorithm" is derived from his name.

Solving Diophantine puzzles requires great acumen and perseverance. Alcuin certainly knew this, and he provided Charlemagne with long hours of intellectual recreation and medieval learners with practice in advanced accurate reckoning. Alcuin's anthology became widely known, and many of his puzzles continue to find their way, in one version or other, into contemporary collections.

The popularity of the *Problems* was matched by several other compilations. One of these, a collection of a hundred mechanical puzzles titled *The Book of Ingenious Devices*, was put together by the eighth-century inventor Mohammed ibn Musá ibn Shakir of Baghdad. Another, titled *The Book of Precious Things in the Art of Reckoning*, was written by the ninth-century Egyptian mathematician Abu Kamil. By the thirteenth century, such anthologies had become commonplace. Of those that have come down to us, Fibonacci's *Book of the Abacus*, published in 1202, is undoubtedly the most significant. Fibonacci's real name was Leonardo Pisano, but, as was common in that era, he was better known (and is known today) by his nickname, which is an acronym for *figlio Bonacci*, "son of the Bonacci family." He also sometimes used another nickname, Bigollo, which means "bungling traveler."

Fibonacci designed his book as a practical, user-friendly introduction to the Hindu-Arabic number system, which he had learned to use during his extensive travels. Incidentally, in it Fibonacci solved the intellectual puzzle of the *zero* concept that mystified the philosophers of his era. The 0 symbol probably originated as far back as 600 B.C. in India, although a symbol for this concept existed in other parts of the ancient world at different times (Kaplan 1998). It was Al-Khwarizmi who introduced the symbol 0, which he called *as-sifr*, "number emptiness" (a translation of the Hindu word *sunya*, meaning "void" or "empty"), to Europe. If it stood for "nothing," philosophers of the era argued, then it surely was "nothing," and thus had no conceivable uses. Fibonacci solved their dilemma by showing that 0 did indeed have a very practical function. It was, he claimed, no more and no less than a convenient arithmetical sign—a digital place-holder for separating columns of figures. Zero is needed in a positional numeral system because the place a digit occupies determines its value. Fibonacci showed this by taking a numeral such as 547. In this numeral the digit 5 stands for 5×10^2, or 5×100 (= 500); the digit 4 for 4×10^1, or 4×10 (= 40); and the digit 7 for 7×10^0, or 7×1 (= 7). The numeral 547 thus represents the number "five hundred plus forty plus seven," or "five hundred and forty-seven" for short. Now, in such a system, how would one write a number such as "five hundred and six," Fibonacci asked? Clearly, some place-filling symbol was needed to show that there were no tens. That symbol is 0, Fibonacci asserted. The appropriate numeral is therefore 506, which is read as "five hundred plus no tens plus six," or "five hundred and six," by convention.

The puzzles Fibonacci devised for his *Book of the Abacus* not only exemplified fundamental principles of mathematics but, like those of Ahmes and Alcuin before him, also made the study of mathematics pleasurable to a large audience. However, unlike those of his two predecessors, Fibonacci's puzzles often concealed traps or twists. Solving his puzzles requires a large dose of *insight thinking,* rather than straightforward accurate reckoning. Here is an example of one of his tricky contrivances. It is found in different versions to this day in many standard puzzle collections:

> *A snake is at the bottom of a 30-foot well. Each day it crawls up 3 feet and slips back 2 feet. At that rate, when will the snake be able to reach the top of the well?*

An inadvertent solver might be induced by Fibonacci's puzzle to reason somewhat as follows. At the end of the first day, the snake will have gone up 1 foot from the bottom of the well (3 feet up and 2 down). At the start of the second day, the snake will begin its journey from the 1-foot level, facing 29 more feet to get to the top (since the well is 30 feet deep). During that day, it will ascend to the 4-foot level, and then slide down to the 2-foot level. So, at the start of the third day, the snake will begin on the 2-foot level, facing 28 more feet to get to the top. During that day, it will ascend to the 5-foot level and then slide down to the 3-foot level. At the start of the fourth day, therefore, the snake will take off from the 3-foot level, facing 27 more feet. To avoid repeating such monotonous calculations, our inadvertent solver might, at this point, be tempted to generalize that the snake's upward crawling rate is 1 foot per day and, consequently, that the snake will get to the top of the well on the twenty-ninth day, since that is the day when it would start on the 28-foot level, crawl up 2 feet to the top, and slither out before sliding back down. But this is not the solution. Consider the snake's movements on the twenty-eighth day. At the start of the day, it begins its journey on the 27-foot level—the level it had reached, of course, at the end of the twenty-seventh day. It goes up 3 feet to the top, where it can crawl out, before sliding. End of matter! Those 3 feet up on the twenty-eighth day are enough to take the snake right up to the top of the well.

The *Book of the Abacus* did not just provide intellectual fun and games. It also helped to establish the Hindu-Arabic numeral system throughout Europe. One of its puzzles, moreover, has had far-reaching implications, since its solution turns up in many ways as a pattern in Nature, as well as in art and many other areas of human invention. Known today as Fibonacci's Rabbit Puzzle, it is found in the third section of the *Book:*

> *A certain man put a pair of rabbits, male and female, in a very large cage. How many pairs of rabbits can be produced in that cage in a year if every month each pair produces a new pair which, from the second month of its existence on, also is productive?*

Note that there is 1 pair of rabbits in the cage at the start. At the end of the first month, there is still only 1 pair, for the puzzle states that a pair is productive only "from the second month of its existence on." It is during the second month that the original pair will produce its first offspring pair. Thus, at the end of the second month, a total of 2 pairs, the original one and its first offspring pair, are in the cage. Now, during the third month, only the original pair generates another new pair. The first offspring pair must wait a month before it becomes productive. So, at the end of the third month, there are 3 pairs in total in the cage: the initial pair, and the two offspring pairs that the original pair has thus far produced. If we keep tabs on the situation month by month, we can show the sequence of pairs that the cage successively contains as follows: 1, 1, 2, 3. The first digit represents the number of pairs in the cage at the start; the second, the number after one month; the third, the number after two months; and the fourth, the number after three months.

Now, consider what happens during the fourth month. The original pair produces yet another pair; so too does the first offspring pair. The other pair has not started producing yet. Therefore, during that month, a total of 2 newborn pairs of rabbits are added to the cage. Altogether, at the end of the month there are the previous 3 pairs plus the 2 new-born ones, making a total of 5 pairs in the cage. This number can now be added to our sequence: 1, 1, 2, 3, 5. During the fifth month, the original pair produces another newborn pair; the first offspring pair produces a pair as well; and now the second offspring pair produces its own first pair. The other rabbit pairs in the cage have not started producing offspring yet. So, at the end of the fifth month, 3 newborn pairs have been added to the 5 pairs that were previously in the cage, making the total number of pairs in it 5 + 3 = 8. We can now add this number to our growing sequence: 1, 1, 2, 3, 5, 8. Continuing to reason in this way, we can show that after twelve months, there are 233 pairs in the cage:

After How Long?	How Many Pairs in the Cage?
the start	1 pair
one month	1 pair
two months	2 pairs
three months	3 pairs
four months	5 pairs
five months	8 pairs
six months	13 pairs
seven months	21 pairs
eight months	34 pairs
nine months	55 pairs
ten months	89 pairs
eleven months	144 pairs
twelve months	233 pairs

The number of pairs in the cage each month can be represented with the following sequence:

1, 1, 2, 3, 5, 8, 13, 21, 34, 55, 89, 144, 233

The salient characteristic of this sequence is that each number in it is the sum of the previous two: e.g., 2 (the third number) = 1 + 1 (the sum of the previous two); 3 (the fourth number) = 1 + 2 (the sum of the previous two); etc. Known aptly as the *Fibonacci sequence*, it can of course be extended ad infinitum, by applying the simple rule of continually adding the two previous numbers to generate the next:

1, 1, 2, 3, 5, 8, 13, 21, 34, 55, 89, 144, 233, 377, 610, 987, . . .

Little did Fibonacci know how significant his simple puzzle would become. For the present purposes, suffice it to say that it has been found to conceal many unexpected mathematical patterns. For example, if the nth number in the sequence is x, then every nth number after x turns out to be a multiple of x:

- the third number is 2, and every third number after 2 (8, 34, 144, . . .) is a multiple of 2;
- the fourth number is 3, and every fourth number after 3 (21, 144, 987, . . .) is a multiple of 3;
- the fifth number is 5, and every fifth number after 5 (55, 610, . . .) is a multiple of 5;

and so on.

As will be discussed later in this book, for some "obscure" or "secret" reason, as Ahmes would have phrased it, this series also surfaces unexpectedly in Nature and in human activities: e.g., daisies tend to have 21, 34, 55, or 89 petals (= the eighth, ninth, tenth, and eleventh numbers in the sequence); trilliums, wild roses, bloodroots, columbines, lilies, and irises also tend to have petals in Fibonacci numbers; a major chord in Western music is made up of the octave, third, and fifth tones of the scale, i.e., of tones 3, 5, and 8 (another short stretch of consecutive numbers in the Fibonacci sequence). The list of such *reifications* is truly astounding. "Reification" is the term used by philosophers to refer to serendipitous actual manifestations of something that was originally conceived as an abstraction or as a figment of mind.

Why would the solution to a simple puzzle reveal patterns in the real world? There is, to the best of my knowledge, no definitive answer to this question; nor, probably, can there ever be one. As mathematician Morris Kline (1985: 42) aptly notes, not only are we "completely ignorant about the underlying reasons" for such coincidences, but "we shall perhaps always remain ignorant of them." All that can be said, in the context of the present discussion, is that Ahmes' characterization of puzzle-solving as a means of gaining "entrance into the knowledge of all existing things and all obscure secrets" turns out to be prophetic, rather than just an exercise of poetic license. Because of its mysterious properties, the Fibonacci sequence has generated enormous interest—so much so that a Fibonacci Society was founded in 1962, and its journal, *The Fibonacci Quarterly,* was first published in 1963.

Like Alcuin's *Problems,* Fibonacci's *Book of the Abacus* was used as a textbook throughout Europe. The facility with which such puzzle books clarified complex mathematical issues and brought out the real-world implications of mathematics encouraged mathematicians of the era to adopt puzzle-making as part of a serious approach to the systematic study of their discipline. The Arab mathematician Ibn Kallikan (c. 1256), for instance, devised a famous puzzle that he used to illustrate the nature of *geometric sequences.* One such sequence is $\{1, 2, 4, 8, 16, 32, 64, 128, \ldots\}$. Each term in it can be expressed as a power of 2, as follows: $1 = 2^0, 2 = 2^1, 4 = 2^2, 8 = 2^3$, etc. The *general term* of this progression is, therefore, 2^n, where $n = \{0, 1, 2, 3, 4, \ldots\}$. Kallikan's puzzle is paraphrased below:

> *How many grains of wheat are needed on the last square of a 64-square chessboard if 1 grain is to be put on the first square of the*

board, 2 on the second, 4 on the third, 8 on the fourth, and so on in this fashion?

If one grain ($= 2^0$) of wheat is put on the first square, two grains ($= 2^1$) on the second, four on the third ($= 2^2$), eight on the fourth ($= 2^3$), and so on, it is obvious that 2^{63} grains will have to be placed on the sixty-fourth square. Kallikan's puzzle illustrates, in effect, the practical meaning of the geometric progression with general term 2^n. The term 2^{63} is the sixty-fourth term in the sequence. Now, Ibn Kallikan asserted, the value of 2^{63} is so large—as readers can confirm for themselves if they so wish—that it boggles the mind to think of what kind of chessboard could hold so many grains, not to mention where so much wheat could be found. Clearly, the goal of Ibn Kallikan's Grain Puzzle was to highlight the enormity of geometric exponential growth. Incidentally, it has been found that all kinds of geometric sequences manifest themselves as patterns in Nature. The French mathematician Jean Baptiste Fourier (1768–1830), for instance, discovered the occurrence of a *trigonometric sequence* in the structure of waves. Again, such fortuitous reifications in Nature of a pattern devised by humans not only give more and more weight to Ahmes' assertion that mathematical knowledge mysteriously provides "entrance into the knowledge of all existing things," but also highlight the fecundity of the puzzle instinct in the human search for understanding.

By the fifteenth and sixteenth centuries, puzzle-making had become a lucrative craft in its own right. In that era certain puzzles were thought to possess occult powers and secret aesthetic qualities. *Magic squares,* for instance, were carried on amulets and talismans to ward off evil. Originating in China around 2400 B.C., magic squares are square arrays of numbers in which the rows, columns, and major diagonals all have the same sum (chapter 5). The great German painter Albrecht Dürer (1471–1528) included a magic square in his famous 1514 engraving *Melancholia.* Nearly two centuries after that, the Swiss mathematician Leonhard Euler (1707–1783) became so mesmerized by Dürer's square that he constructed forty-eight versions of the square himself.

With the invention of movable metal-type printing in the fifteenth century, making books widely available, more and more puzzle collections were published. Well-known mathematicians, such as the Welshman Robert Recorde (c. 1510–1558), who invented the equal sign "=", and the Italians Niccoló Fontana Tartaglia (c. 1499–1557) and Girolamo Cardano (1501–1576), who jointly discovered the algebraic solution to cubic equations (equations in which the greatest exponent on a variable is 3), increasingly employed the puzzle format to illustrate mathematical concepts. By the seventeenth century, at the dawn of the modern period, puzzle compilations had become widely accepted both as materials for delectation and as tools for illustrating and probing mathematical ideas.

In 1612, the French Jesuit poet and scholar Claude-Gaspar Bachet de Mézirac (1581–1638) published what has become one of the all-time best-selling collections of puzzles, titled *Amusing and Delightful Number Problems,* a book that continues to provide many amusing and delightful moments to this day. Coincidentally, Bachet also translated, in 1621, Diophantus's *Arithmetica,* the book Pierre de Fermat (1601–1665) was reading when he inscribed his famous comment about Pythagorean number triples in its margin (chapter 7). In his collection, Bachet put forward the first-ever classification of puzzles—river-crossing puzzles, weighing puzzles, number tricks, etc.—a system that has been adopted *grosso modo* by puzzlists ever since. Another widely known puzzle anthology of the same era was Henry van Etten's *Mathematical Recreations; or, A Collection of Sundrie Excellent Problemes out of Ancient and Modern Phylosophers Both Usefull and Recreative,* published in French in 1624 and then in English in 1633. Van Etten took freely from the work of his predecessors, especially from Bachet and the *Greek Anthology,* but he also introduced many innovative puzzles of his own. Shortly thereafter, in 1647, the first collection of puzzles in America was published in an almanac printed by Samuel Danforth, an emigré from England. Among its many interesting items, the almanac contains twelve versified riddles.

In the eighteenth century, more and more mathematicians created puzzles as a means of stimulating broad interest in new mathematical ideas. With his Thirty-Six Officers Puzzle of 1779, Euler, for instance, was able to arouse great interest in what was then a fledgling and uninviting area of mathematics, subsequently named *combinatorics.* The puzzle asks, simply, if it is possible to arrange 6 regiments consisting of 6 officers, each of different rank, in a 6 × 6 square so that no rank or regiment will be repeated in any row or column. Such an arrangement turns out to be impossible, but the puzzle quickly became one of the cornerstones in the edifice of combinatorics that was being built at the time. Euler's best-known puzzle is, however, his Königsberg's Bridges Puzzle:

> *In the town of Königsberg, is it possible to cross each of its seven bridges over the Pregel River, which connect two islands and the mainland, without crossing over any bridge twice?*

The puzzle can be represented visually as follows. The double lines stand for the bridges of Königsberg; A and B are the two islands that are connected to the city by the bridges, and C and D are the mainland areas:

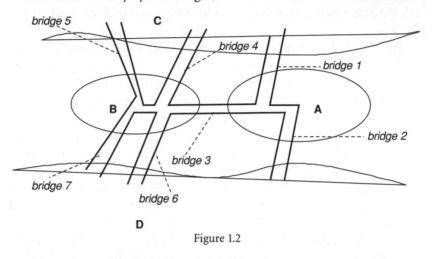

Figure 1.2

Euler showed that it is impossible to trace a path over the bridges without crossing at least one of them twice. He started his demonstration by rephrasing the puzzle as follows:

> *Is it possible to draw the following graph without lifting pencil from paper, and without tracing any edge twice?*

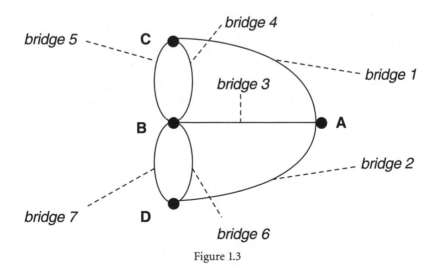

Figure 1.3

Euler's graph—called a *network* in contemporary combinatorics—simply redraws the map in outline form, showing the land areas as *points* (vertices) and the bridges as *paths* (lines and arcs) between them. In the language of combinatorics, an *even point* is one at which an even number of paths converge, and an *odd point* is one at which an odd number of paths converge. In order to grasp Euler's solution, it is helpful to consider a few simple networks with different numbers of even and odd points (Pappas 1989):

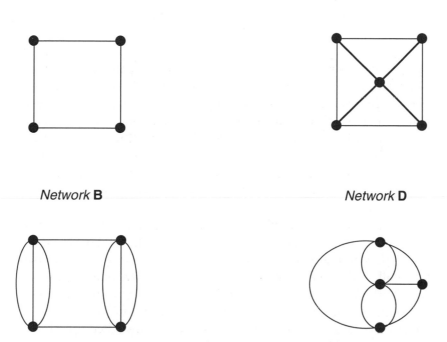

Figure 1.4

In network A, an even number of linear paths (two) converge at each of its four points. By putting a pencil on any dot and tracing the paths, the reader can traverse this network without having to double back over any path. In network B, an even number of paths (four, two linear and two curved) similarly converge at each of its four points. Again, by tracing the network with a pencil, the reader will be able to see that this one, too, can be traversed without having to double back over a path already traced. In network C, however, an odd number of linear paths (three) meet at each of the four outer points, and an even number (four) at the inner point. Tracing this network without doubling back turns out to be impossible. In network D, the top point is even, because four curved paths converge there; the one below it is odd, because one linear and four curved paths meet there; the bottom point is even; and the point to the right is odd. In total, there are two odd and two even points. Network D can be traversed without doubling back over any path (as readers can verify for themselves).

Creating other simple networks in this way, with more and more paths and points in them, will show that it is not possible to traverse a network with more than two odd points in it without having to double back over some of its paths. Euler proved this tentative conclusion with a remarkably simple argument. It goes somewhat like this. In arriving at a point with a pencil and then leaving it, two of the paths are "used up." Therefore, any point at which an odd number of paths meet (all of which must, of course, be traversed) can only be a starting point or an ending point; otherwise there would be a path "left over" in the network. Now, the number of paths meeting at each point of the Königsberg network [see figure 1.3] are A = 3, B = 5, C = 3, D = 3. These are all odd. Consequently, Euler concluded, it cannot be traced by one continuous stroke of a pencil without having to double back over paths that have already been traced.

Several decades after Euler's demonstration, mathematicians began studying networks seriously. Their efforts led to the establishment of a new branch of mathematics called *topology,* which explores the properties of all kinds of networks. The first comprehensive treatise of topology, titled *Theory of Elementary Relationships,* was published in 1863. It was written by the German mathematician Augustus Möbius (1790–1868), the inventor of a truly enigmatic topological figure called the Möbius Strip, which will be discussed in chapter 3.

The nineteenth century saw a growth of interest in combinatorial mathematics. In that century another famous combinatorial puzzle, known as Kirkman's School Girl Puzzle—named after the notable amateur mathematician Thomas Penyngton Kirkman, who posed it in 1847— had important implications for matrix theory. A *matrix* is a rectangular array of numerical or algebraic symbols arranged in columns and rows:

> *How can 15 girls walk in 5 rows of 3 each for 7 days so that no girl walks with any other girl in the same triplet more than once?*

Solutions to this puzzle involve arranging the fifteen numerals from 0 to 14 (each one representing a specific girl) in five rows of three each within seven sets (each set corresponding to a day of the week) so that no two numerals appear in the same row more than once (see solution 1.3). Since 1922, various solutions to Kirkman's problem have been found (Gardner 1997: 125–126). Complex puzzles such as this one were of major interest to mathematicians and puzzlists throughout the century. For example, the British mathematician and logician Augustus De Morgan (1806–1871), who wrote important works on the calculus and modern symbolic logic, produced a truly ingenious collection of puzzles, titled *A Budget of Paradoxes,* in which he explored a host of mathematical theories, ideas, and suppositions that were being bandied about at the time.

Whereas mathematicians saw the puzzle format as a means of presenting new theoretical notions in a recreational vein, others saw it instead as something quite different—as a new literary genre. Many famous writers and philosophers of the era also wrote puzzles with the same creativity that they brought to their literary pursuits. The *Ladies' Diary,* or *Woman's Almanac* reflected this new perception of puzzles perfectly. Founded in 1704 and initially consisting of recipes, portraits of notable women, and articles on health and education, it soon became the first puzzle magazine of history, providing its readers with poetically written puzzles of all kinds for intellectual amusement. The magazine ended publication in 1841.

It was the English writer and mathematician Lewis Carroll who raised the puzzle genre to a high literary art. Lewis Carroll was the *nom de plume* of Charles Lutwidge Dodgson. He created the first puzzle storybooks of history with the publication of *Pillow Problems* in 1880 (a compilation of seventy-two puzzles in arithmetic, algebra, geometry, trigonometry, calculus, and probability) and *A Tangled Tale* in 1885 (a collection of puzzles originally published in monthly magazine articles). Carroll is best known, of course, for his immortal fantasies *Alice's Adventures in Wonderland* (1865) and *Through the Looking-Glass* (1871). But he was also an eminent mathematical theorist, writing two famous treatises, *A Syllabus of Plane Algebraical Geometry* (1860) and *Euclid and His Modern Rivals* (1879), both of which were widely read by the mathematicians of the era. Carroll was fascinated by the inquisitive and fanciful imagination of children. *Alice's Adventures in Wonderland* contains all sorts of puzzles involving ingenious mind play and double-entendre that have amused and challenged children ever since the book was first published. Carroll was obviously captivated by the ability of puzzles to impose a peculiar kind of ordered thinking on the erratic and capricious human mind. As Cohen (1995) has persuasively written, this probably allowed Carroll to cathartically exorcise his emotional agony. As Carroll obviously understood, finding solutions to puzzles provides reassurance and a sense of order. As the British mystery writer P. D. James (b. 1920) so aptly put it in a magazine interview (James 1986), mystery stories and puzzles are really all about the "restoration of Order." Like works of art, puzzles appear to have no other function than to provide our need for order a channel through which to express itself.

The nineteenth century saw puzzle-making become increasingly popular as a distinct genre in the print medium. Collections such as the *Rational Amusements for Winter Evenings* by John Jackson (1821) and the *Recreations in Mathematics and Natural Philosophy* by Edward Riddle (1840) became widely known. The latter work was a revision of a puzzle collection compiled by Jacques Ozanam (c. 1640) and later enlarged by Jean Etienne Montucla (1725–1799). The Frenchman François Edouard Anatole Lucas (1842–1891), well known for his writings on mathematics, became widely known as the inventor of the Towers of Hanoi Puzzle, which he fashioned after one of Girolamo Cardano's puzzle creations. Perhaps to keep his reputation as a mathematician distinct from his interest in puzzle-making, he published his Towers puzzle in 1883 under the pseudonym M. Claus de Siam, "Claus" being a rearrangement of the letters of "Lucas." The puzzle is paraphrased below:

> *A monastery in Hanoi has three pegs. One holds 64 gold discs in descending order of size—the largest at the bottom, the smallest at the top. The monks have orders from God to move all the discs to the third peg while keeping them in descending order. A larger disc must never sit on a smaller one. All three pegs can be used. When the monks move the last disk, the world will end. Why?*

As we shall see in chapter 6, the reason the world will end is that it will take the monks $2^{64} - 1$ moves to accomplish the task God set before them. Even at one move per second (and no mistakes), this task will require 582,000,000,000 years!

By century's end, the first "professional puzzlists" had come onto the social scene—the American Sam Loyd (1841–1911) and the Englishman Henry E. Dudeney (1847–1930). Sam Loyd was the first individual to earn a truly comfortable living from puzzle-making. As Matthew J. Costello (1996: 45) describes him, Loyd was "puzzledom's greatest celebrity, a combination of huckster, popularizer, genius, and fast-talking snake oil salesman." For the cleverness and sheer volume of his work, many consider Loyd to be the greatest puzzlist of all time. His love for puzzles apparently started in 1860, when he became problem editor of the magazine *Chess Monthly*. In 1878, Loyd put together a compilation of chess puzzles, titled *Chess Strategy*. So successful were both ventures that he became convinced he could make a decent living working only as a puzzlist, even though he was trained as an engineer. Working out of a small, dusty office in Manhattan, Loyd produced over ten thousand puzzles in his lifetime, most of which were extremely challenging, forcing puzzle addicts to spend countless hours trying to figure them out.

His most famous and lucrative puzzle creation was the 14/15 Puzzle, which he put out in 1878. Loyd's seemingly trivial gadget produced a worldwide "craze." In America, employers felt it necessary to put up notices prohibiting their employees from working on the puzzle during office hours. In France, it was characterized as a greater scourge than alcohol or tobacco. Loyd's gadget was a slim square case capable of holding 16 small square sliding blocks but containing, instead, 15 blocks, numbered from 1 to 15. The blocks were arranged in order from 1 to 13. The two numbered 14 and 15, however, were reversed. The challenge was to arrange all the blocks in order from 1 to 15 without removing any of them physically from the case: i.e., by sliding one block at a time into the empty square. As the roguish Loyd certainly knew, the puzzle cannot be solved. This nonetheless intriguing gadget will be discussed in chapter 6.

Henry Ernest Dudeney was Loyd's British counterpart. At age nine, Dudeney started inventing mind-bending puzzles that he published in a local paper, under the appropriate pseudonym of Sphinx. In 1893, Loyd and Dudeney started a correspondence, as leading puzzlists of the day. But Dudeney eventually became upset with his American pen pal, breaking off relations after he started suspecting Loyd of plagiarizing his puzzle ideas. Dudeney contributed to the *Strand Magazine* for over thirty years, and he wrote a number of challenging puzzle books that have remained highly popular to this day. His 1917 masterpiece of enigmatology, *Amusements in Mathematics,* has been the model for all puzzle anthology designs ever since.

The twentieth century witnessed a massive proliferation of interest in puzzles of all kinds. It also saw the invention of the *crossword puzzle* in 1913 and the Rubik's Cube in 1975, both of which became instant crazes, as well as the establishment of a plethora of puzzle leagues, tournaments, and specialized magazines. Puzzle associations were also founded, such as the Association of Game and Puzzle Collectors in 1985 and the Slocum Puzzle Foundation in 1993. Jerry Slocum, incidentally, has also put together several interesting compilations of puzzles and games with co-author Jack Botermans (1992, 1994). Of the magazines, *Games* was founded in 1977, *The Cryptogram* (founded in 1932) is published by the American Cryptogram Association, and *The Enigma* (founded in 1883) is published by the National Puzzlers' League. *The Cryptogram* has a circulation of about 400, and *Games* of about 600. Puzzle toys and games designed for children also became so popular that some manufacturers decided to specialize in such toys. And, since the late 1980s, the number of Web sites about puzzles and games has become truly astronomical. Incidentally, the first "cyber-puzzles" were created in the mid-1970s by American computer programmers Willie Crowther and Don Woods, who distributed them on the computer network ARPAnet, the predecessor to today's Internet.

Puzzle-making as a profession was also pursued by a growing number of individuals. Mathematician W. W. Rouse Ball, for instance, came to be widely known for his extensive and in-depth treatment of famous puzzles, *Mathematical Recreations and Essays,* which was published in 1892 and reissued in at least thirteen more editions over the next century. Two other widely known puzzle authors, Martin Gardner (b. 1914) and Raymond Smullyan (b. 1919), made puzzle-making a profession. Gardner has written extensively on mathematical puzzles, making it his lifetime goal to delve into the history and meaning of puzzles (a list of Gardner's writings is included in the bibliography at the back of this book; a study of his work can be found in Richards 1999). He wrote a puzzle column for *Scientific American* for nearly thirty years (starting in December 1956), where he popularized many of the puzzles of Loyd and Dudeney, along with those of contemporary puzzlists. As Klarner (1981: iii) puts it, in the domain of puzzle-making Gardner was "a force behind the scenes as well as a public figure." Smullyan is a mathematical logician and has composed a host of ingenious chess and logic puzzles designed to strip down logical reasoning to its bare essentials. He devised his first puzzles in 1935, when he was barely sixteen years old. Smullyan's books are delectable and, in the tradition of the nineteenth-century literary puzzlists, readable as narrative works in their own right (a partial list of his works is included in the bibliography).

Puzzle-Solving as Insight Thinking

The word "puzzle" is probably derived from the Middle English word *poselet*, "bewildered, confused." It is an apt term because, unlike the typical problems found in textbooks, which are designed to test knowledge of specific mathematical principles, puzzles generate bewilderment and confusion. They do so by concealing either a pattern or a twist or trap within the information they present. Euler's bridges puzzle is an example of the former, and Fibonacci's snake puzzle of the latter. As Helene Hovanec (1978: 10) has aptly stated in her delightful book, *The Puzzler's Paradise*, the lure of puzzles lies in the fact that they "simultaneously conceal the answers yet cry out to be solved," piquing solvers to pit "their own ingenuity against that of the constructors."

Puzzles are not solved by the use of accurate reckoning alone (to use Ahmes' phraseology), but also (and above all else) by a substantial use of insight thinking. Consider, as a case in point, the following classic puzzle, which shows initially an arrangement of six sticks laid out on a table representing the fraction ½ in Roman numerals (I/VII):

Figure 1.5

By moving a single stick, other than the horizontal one, is it possible to change the above arrangement to one showing a fraction equal to one?

The solution to this puzzle can hardly be discovered by accurate reckoning. Only by envisioning (i.e., mentally picturing) possible arrangements of the sticks can it be found. The solution involves joining the V figure with one of the two upright sticks in the denominator to make a square root sign. This new arrangement shows the numeral I over √I (= 1); and the resulting fraction is, of course, equal to I:

$$\frac{|}{\sqrt{|}}$$

Figure 1.6

As spontaneously as it may seem to have come to mind, the above solution is hardly disconnected from previous experience and knowledge. Only someone who is familiar with square root signs and who has reflected upon the various patterns that constitute numerical forms can envision the solution. In a phrase, insight thinking does not emerge fortuitously or haphazardly. It comes about only after the observation and contemplation of recurring patterns. Insight thinking can be defined as the ability to see with the mind's eye the inner nature of some specific thing. The psychologist Robert Sternberg (1985) argues, in fact, that insight thinking is anchored in three forms of reflective memory: (1) *selective encoding,* or the use of information that may have originally seemed irrelevant but that may become crucial in due course; (2) *selective comparison,* or the discovery, often through analogical and metaphorical thinking, of a nonobvious relationship between new information and information already in memory; and (3) *selective combination,* or the discovery of nonobvious pieces of information that can be combined to form novel information and ideas. This *triarchic* process, as Sternberg calls it, is certainly the source of the insight that was required to solve the above puzzle.

The great American mathematician and philosopher Charles Sanders Peirce (1839–1914) characterized an *insight* as an "informed hunch," based on previous experience, as to what something entails or presupposes. He called the thinking process that produces it *abductive.* A classic example of abductive thinking in the domain of physics is the model of atomic structure originated by the English physicist Ernest Rutherford (1871–1937). Rutherford guessed that the inside of an atom had the structure of an infinitesimal solar system, with electrons behaving like planets orbiting an atomic nucleus. In effect, Rutherford envisioned the atom as a tiny solar system. His insight was extended shortly thereafter by the great Danish physicist Neils Bohr (1885–1962). Rutherford's model, in which electrons move around a tightly packed, positively charged nucleus, successfully explained the results of scattering experiments, but was unable to explain atomic emission (why atoms emit only certain wavelengths of light). Bohr reasoned further that electrons can only move in certain "quantized" orbits. His new insight was thus able to explain certain qualities of emission for hydrogen, but it failed to account for those of other elements. The Austrian theoretical physicist Erwin Schrödinger (1887–1961) then incorporated Rutherford's and Bohr's work into his own, in order to describe electrons not by the paths they take but by the regions where they are most likely to be found. This insight allowed him to explain certain qualities of emission spectra for all elements.

Now, consider the following puzzle:

> *A man was watching his son pick apples, noticing that the number of apples in his basket doubled every minute and that it was full at precisely 12 noon. At what time was the basket half full?*

The insight in this case involves thinking backward from noon to 11:59. Let us assume, for the sake of concreteness, that a full basket contains 20 apples. That is what the basket had in it at 12:00 noon. Since we are told that the number of apples doubled every minute, we can easily *envision* that at 11:59 there were 10 apples in the basket. It was during the next minute that the son doubled the basket's content to 20. That is, in fact, the solution—the basket was half full at 11:59.

Sometimes a form of envisioning that may be called *mental projection* is involved. Consider the following puzzle as a case in point:

> *Sally and Peter both want to catch the 8:00 flight to Chicago. Sally thinks that her watch is 25 minutes fast, although it is actually 10 minutes slow. Peter thinks his watch is 10 minutes slow, although it is actually 5 minutes fast. If both persons rely on their watches, what will happen if they attempt to arrive at the airport 5 minutes before the flight departs?*

In this case, the solution hinges on projecting oneself into the minds of the characters of the puzzle, i.e., on envisioning what they see and think as they look at their watches. When Sally looks at her watch and sees 8:20, she will think that it is 7:55, because she believes that her watch is 25 minutes fast. In reality, her watch is 10 minutes slow, and the time will be 8:30. Since the plane is scheduled to leave at 8:00, Sally will miss her flight. Now, when Peter looks at his watch and sees 7:45, he will also think that it is really 7:55, because he believes that his watch is 10 minutes slow. Therefore Peter, like Sally, will think he has 5 minutes to spare before takeoff time at 8:00. In reality, since his watch is 5 minutes fast, the time will actually be 7:40. So Peter will arrive ahead of departure time at 8:00.

This puzzle further highlights the role of previous experience in insight thinking. The solution is envisionable only by someone who is familiar with, and has reflected upon, how clocks mark the passage of time. The following puzzle, too, which is a version of a problem devised by Jacques Ozanam (mentioned above), cannot be solved by anyone who has not thought reflectively and deliberately about how clocks mark time. The solution requires envisioning the movement of the hands of a clock past the divisions marked on its face. Its solution is left as an exercise for the reader (see solution 1.4):

> *How many minutes after 8:00 will the minute hand overtake the hour hand?*

Insight thinking can be characterized, in sum, as an admixture of imagination and memory that leads us to literally see the pattern or twist that a puzzle conceals. It is, in a sense, clairvoyance, since it entails perceiving things that are not immediately evident. Incidentally, comparing puzzle-solving to clairvoyance suggests a plausible reason why so many puzzles tend, in their origins, to be associated with occult or mystical thinking. Finding the solution to a difficult puzzle through a "flash of insight" is perceived even today—as it certainly must have been in the ancient world—as mysteriously revelatory.

Insight thinking is well known among puzzlists. In his intriguing book *Aha! Insight!* Martin Gardner (1979a) takes us on his own fascinating journey through Puzzleland in order to impress upon us that no extensive mathematical training is required to solve puzzles, since solutions tend to come suddenly, by flashes of insight. This is why puzzles often provoke an *Aha!* or *Eureka!* effect—something similar to the notion of mental catharsis mentioned at the start of this chapter. These exclamations express relief, pleasure, and triumph all at once. "Eureka," which means "I have found it" in Greek, is what Archimedes supposedly shouted when he envisioned a way to determine the purity of gold by applying the principle of specific gravity.

Insight thinking underlies all mathematical and scientific discoveries. As the late historian of science Jacob Bronowski (1977: 24) eloquently put it, the very process of reasoning in mathematics and science relies upon the ability of the human imagination "to make images and to move them about inside one's head in new arrangements," because "unless connections between things are seen in the mind, there is nothing to reason about." Instances of insight thinking, in fact, pervade the annals of science. Take, for example, the calculation of the earth's circumference by the Greek astronomer, geometer, and geographer Eratosthenes of Alexandria (c. 275–194 B.C.). Standing at Alexandria during the summer solstice, and knowing that it was five hundred miles due north of the city of Syene, Eratosthenes had a flash of insight into how to measure the earth's circumference—without having to do it physically. He knew, as an astronomer, that at the summer solstice the noon sun was directly overhead at Syene. So he drew a diagram showing the earth as a circle, labeling its center with the letter O and the cities of Alexandria and Syene with A and B, respectively, as shown in the figure below:

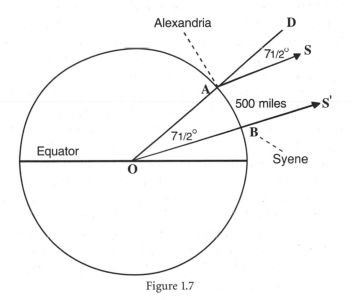

Figure 1.7

He then represented the direction of the sun (S') over Syene as BS', and he envisioned that joining this line to the earth's center O would form the straight line OBS', since at that instant the sun would be shining directly overhead. At the same moment in Alexandria, he argued, the direction of the sun (S) could be represented with AS, a line parallel to BS', because, as he knew, the sun's rays are parallel to each other. Then he extended the line OA—the line joining the center of the earth to Alexandria—to an arbitrary point D directly above Alexandria. Now, Eratosthenes reasoned, since AS was parallel to BS', the angles ∠DAS and ∠AOB, formed by the line OAD cutting the two parallel lines AS and OS', were equal (according to a theorem of Euclidean geometry). On the basis of this knowledge, Eratosthenes had only to calculate ∠DAS, which he was able to do easily, being in Alexandria, by measuring the angle of the shadow made by a well near which he was standing. This was the difference between the overhead direction and the direction of the sun at Alexandria (and thus at Syene). He found it to be 7½°. This, then, was also the size of ∠AOB. Moreover, Eratosthenes reasoned, this angle is 7½° of 360°, since the earth is virtually a sphere. Thus, ∠AOB was 1/48 of the entire angle at O (7½° of 360°). From another theorem of geometry, Eratosthenes knew that the arc AB, the distance between Alexandria and Syene, was therefore 1/48 of the circumference. Thus, Eratosthenes concluded, the circumference was 48 times the length of that arc: 48 × 500 = 24,000 miles. His calculation of 24,000 miles was in close agreement with the modern value of 24,844 miles.

This often-cited episode in the history of science shows clearly how powerful insight thinking is. It allows physical intervention to be eliminated, permitting humans to envision the physical world in abstract ways which allow them to discover real properties of that world. But it must not be forgotten that the reason why insightful representations such as the one devised by Eratosthenes produce the results they do is that they are based on previous experience and knowledge.

No wonder, then, that puzzle-solving has been associated since ancient times with intelligence. The Riddle of the Sphinx was a kind of mythical intelligence test, as was the Cretan labyrinth. To this day, in fact, skill at puzzle-solving is thought to be the privilege of those with superior intelligence. But the modern-day notion of measurable intelligence is hardly what the ancients had in mind. In a study published in 1982, the psychologists Sternberg and Davidson investigated the relationship between puzzle-solving and the so-called *intelligence quotient* (IQ) of typical solvers, finding that the capacity to use insights to solve puzzles did not always correlate with IQ. To the best of my knowledge, no study since theirs has come forward to refute or seriously contradict it. Simply put, puzzles are solved by all kinds of individuals, no matter what their IQs are. Of course, some are better at it than others. But this is true of anything in life.

The notion of measurable intelligence is fundamentally flawed. The first intelligence test for standardized use was devised in 1905 by the French psychologist Alfred Binet (1857–1911) and his colleague Théodore

Simon (1873–1961). Binet wanted to ensure that no child would be denied instruction in the Paris school system, no matter what socioeconomic class the child came from. In 1916, the American psychologist Lewis Madison Terman (1877–1956) revised the Binet-Simon test, adapting it specifically to the needs of American society. It was Terman who coined the term "intelligence quotient," believing that the test would provide a scientific basis for comparing the intelligence levels of individuals. But what Terman failed to see was that his test measured only what he himself believed was intelligence. Its validity was therefore called into serious question almost from the moment it was devised. But even before Terman, the notion that intelligence testing could predict success consistently revealed itself to be flawed. Poet Ralph Waldo Emerson (1803–1882), for instance, always scored at the bottom of tests designed to test intelligence as it was conceived during his era; Thomas Edison (1847–1931) was told he was "too stupid" to do anything in life, because of his inability to do well on school tests; Albert Einstein (1879–1955) performed poorly on mathematics tests; and the list of such cases could go on and on.

People of all eras, all cultures, and all kinds of educational backgrounds have made and solved puzzles. Hundreds of thousands of puzzle books are sold every year, and almost every one of the nearly two thousand newspapers published in the United States carries a puzzle. Clearly, puzzle-solving is hardly the exclusive domain of high-IQ achievers. Indeed, sometimes too much "mathematical intelligence" may even get in the way of puzzle-solving. An anecdotal corroboration of this is provided by a legendary story told about the great John von Neumann (1903–1957), the Hungarian-born professor of mathematics at Princeton University, whose ideas were instrumental in the development of the modern computer. The following puzzle was evidently posed to him at a cocktail party:

> *Two children, a boy and a girl, were out riding their bikes yesterday, coming at each other from opposite directions. When they were exactly 20 miles apart, they began racing toward each other. The instant they started, a fly on the handlebar of the girl's bike also started flying toward the boy. As soon as it reached the handlebar of his bike, it turned and started back toward the girl. The fly flew back and forth in this way, from handlebar to handlebar, until the two bicycles met. Each bike moved at a constant speed of 10 miles an hour, and the swifter fly flew at a constant speed of 15 miles an hour. How much distance did the fly cover?*

In one version of the story, von Neumann was apparently so taken by the puzzle that he withdrew to another room to figure out the answer in private. The person who presented the puzzle to the gathering explained its simple solution to the rest of the guests. He started by calculating how much time it took the bike riders to cover the 20 miles between them. Since both were traveling at 10 miles per hour toward each other, they would therefore meet halfway: i.e., after they had each covered 10 of the 20 miles. Moving at 10 miles per hour, the riders covered this distance in 1 hour:

$$D = R \times T$$
$$10 = 10 \times T$$
$$T = 1$$

Since the fly went back and forth at 15 miles per hour during that 1 hour, he stated, it covered a total distance of 15 miles:

$$D = R \times T$$
$$D = 15 \times 1$$
$$D = 15$$

The guests were impressed by the puzzle's ingenuity, since most of them had fallen into the trap of trying to compute the fly's movement back and forth. What they failed to envision was that the fly covered a certain distance, not a certain direction, within the specified period of 1 hour. It does not matter if that distance can be mapped out as a straight line or as a back-and-forth movement.

Von Neumann is said to have come back into the room shortly thereafter and also given the correct answer. The person who posed the puzzle was amazed, remarking that highly skilled mathematicians tended to overlook the simple way in which the puzzle can be solved, trying instead to solve it by a lengthy and complicated process using advanced mathematics (summing an infinite series). Von Neumann is said to have looked quizzically at the fellow, retorting matter-of-factly, "Well, that's precisely how I solved it!"

Rather than call it a form of intelligence, Peirce characterized the insightful form of thinking that reveals itself in the solution of puzzles *logica utens,* which he defined as a rudimentary *logic-in-use* that everyone possesses without being able to specify what it is. He distinguished it from *logica docens,* which he defined as a sophisticated and tutored use of logic practiced by mathematicians, scientists, detectives, and medical doctors (Sebeok and Umiker-Sebeok 1983: 40–41). *Logica utens* is the source of insight thinking; *logica docens* is part of what prepares us to carry out advanced forms of such thinking. Because everyone possesses *logica utens,* no special training is required to understand what puzzles are about or what to do in order to solve them.

The Puzzle Instinct

As suggested in the preface to this book, the intuitive knack for puzzles felt by people across the world can be called an *instinct.* I use this term, of course, in a nonphysical sense. As Keith Devlin cogently argues in his intriguing book *The Math Gene: How Mathematical Thinking Evolved and Why Numbers Are like Gossip* (2000) the word "instinct" is being glibly used nowadays to describe all kinds of abilities, and this usage can be misleading because it assigns to some obscure genetic faculty what may in fact arise from some unknown capacity of the human mind. The thesis espoused in this book is that the appearance of puzzles in human history cannot be easily explained away by some overarching theory of genetic evolution. In the history of humanity, puzzles stand out as truly mysterious artifacts. And for some occult and bewildering reason, the patterns imprinted in some puzzles (such as the Fibonacci rabbit puzzle) are reified in Nature and in human activities. This suggests that insight thinking may itself be a product of forces or energies that are beyond our understanding—forces that interconnect the human mind with Nature.

The puzzle instinct is comparable to what might be called our instinct for humor. No one knows exactly why we are impelled to laugh or why anything that is perceived as funny should cause us to make such a peculiar noise. It would be just as "natural" to do something else, such as stick a finger in our ears, as it is to giggle or howl with laughter. But when something "hits our funny bone," as the expression goes, our diaphragm pulsates up and down, and we laugh. Similarly, when we are given a puzzle to solve, our mind sharpens, our *logica utens* starts working, and we set out to find a solution, as if by instinct, until we are satisfied cathartically.

As Ahmes' prologue to the Rhind Papyrus hints, people in the ancient world sensed deeper, awesome meanings to puzzles. While this sense has become largely unconscious today, it nevertheless still underlies the common feeling of amazement that comes from unraveling the solution to a tricky puzzle—a feeling that was called the *Aha!* or *Eureka!* effect above. This can be called Ahmes' legacy. There is little doubt in my mind that the Rhind Papyrus was intended to intrigue its readers, to generate an *Aha!* effect, and thus to awe them with the power of numbers and mathematical symbols and with the relations among them to provide insights into the nature of things. As Ahmes aptly wrote, puzzle-solving seems to grant us "entrance into the knowledge of all existing things and all obscure secrets." This is why we continue to be intrigued by puzzles of all kinds, as if impelled by some ancient occult force within us to solve them. As Henry Dudeney (cited by Wells 1992: 89) has remarked, "The fact is that our lives are largely spent in solving puzzles; for what is a puzzle, but a perplexing question? And from childhood upwards we are perpetually asking questions and trying to answer them."

Answering the question *Why puzzles?* is akin to answering any other *Why* question: *Why life? Why the universe?* and so on. These are vast questions that surface periodically in our consciousness. They defy answers. On the other hand, answering the questions that specific puzzles pose is much more practicable. Thus, in a fundamental sense, puzzles provide a means of "comic relief," so to speak, from the angst caused by the unanswerable larger questions. As Perkins (2000: 27) aptly puts it, puzzles are "small-scale experiences" of the larger questions that life poses to us. Since there are no definitive answers to the large-scale questions, we are strangely reassured by the answers built into the small-scale ones.

2 Puzzling Language

RIDDLES, ANAGRAMS, AND OTHER VERBAL PERPLEXITIES

"Out of the eater came forth meat, and out of the strong
came forth sweetness." . . . If ye had not plowed with my
heifer, ye had not found out my riddle.

—Judges (14:14, 18)

As the oldest puzzle known, the Riddle of the Sphinx (chapter 1)
is a perfect example of how puzzles tell a fascinating story about human
affairs. The inability to solve the riddle had, as we saw, dire consequences for would-be heroes. While the repercussions of not being able to
solve riddles today are not as catastrophic, it is nonetheless true that
failure to do so leaves a strange, discomforting feeling within us and
makes us feel less than heroic. Children in particular are instinctively
drawn to riddles, clearly enjoying both the challenge they pose and the
mischievous language with which they have been constructed.

The ancients certainly took riddles seriously, as many early legends
attest. The mythical figures of yore earned their heroic status not only
through their physical prowess, but also (if not more) through their ability to solve or invent challenging riddles. The biblical story of Samson is
a case in point. At his wedding feast, Samson wanted to impress the rel-
atives of his wife-to-be by posing the following riddle to his Philistine
guests (Judges 14:14):

*Out of the eater came forth meat, and out of the strong came forth
sweetness.*

He gave the Philistines seven days to come up with the answer, convinced that they would be intellectually incapable of doing so. The clever Samson had devised his riddle as a description of something he had once witnessed—a swarm of bees making honey in the carcass of a lion. The Philistines, however, took advantage of the time given to them to threaten Samson's wife, eventually coercing the answer from her. When they gave Samson the correct answer after seven days, the mighty hero of Hebrew legend was stunned. Enraged, he declared war against the Philistines. The ensuing conflict, during which his Philistine mistress Delilah betrayed him, led eventually to his own destruction. And all this over a simple riddle!

Puzzles based on language structure abound throughout cultures and across time, and they are intrinsically intertwined with the myths, legends, and traditions of all ancient societies. *Anagrams* (words or phrases made by rearranging the letters of given words), for instance, were widely perceived to offer secret messages from the deities that purportedly had the power to predict or affect someone's destiny. It was only after the Renaissance that the widespread view of certain words as magical anagrams started to fade, as Western society began reinterpreting the symbols of the past in new scientific ways. But the feeling that words possess magical qualities was not erased from the human imagination. This feeling can still be seen on children's faces each time they learn a new word or hear a riddle.

The realm of word puzzles is the first stop on our journey through Puzzleland. This is the territory inhabited by Humpty Dumpty—a Carrollian character who is Alice's ironic commentator on the many puzzling features that are built right into the very structure of language, humanity's greatest intellectual achievement.

Riddles

Those who were incapable of solving the Riddle of the Sphinx paid for their ineptitude with their lives. Samson's life ended in calamity over a riddle. To the list of casualties, one can add the ancient Greek poet Homer, whose death was said to have been precipitated by the distress he felt at his failure to solve the following riddle posed to him by a group of fishermen:

What we caught, we threw away. What we could not catch, we kept.

(answer: *fleas*)

For many centuries, riddles were often regarded as coded messages from divine sources, designed to test the intellect and sapience of human beings equipped with special knowledge. This is why, in ancient Greece, priests and priestesses called *oracles* frequently expressed their messages in the form of riddles. They were thought to be extraordinary individuals who had special powers to speak on behalf of the gods and to reveal their will. The use of the word "riddle" in everyday discourse—the "riddle of life," the "riddle of the universe," the "riddle of language," etc.—reverberates with oracular overtones to this very day. If we could figure out such riddles, we seem to feel, we could probably understand the meaning of existence.

Not all the legends associated with riddles were as ominous as those mentioned above. The biblical kings Solomon and Hiram, for example, organized riddle contests simply for the pleasure of outwitting each other. The ancient Greeks used riddling at special banquets as a form of entertainment for the guests, as we might in fact do even today at social gatherings. The ancient Romans made riddles a central feature of the *Saturnalia,* a religious event that they celebrated from December 17 to 23, over the winter solstice. By the fourth century riddles had, in fact, become highly popular for their recreational value all over Europe. The most famous collection in that century, consisting of one hundred riddles in the form of three-line Latin poems, was written by a certain Symphosius, about whom virtually nothing is known. Symphosius gave away the solutions to all his riddles in their titles. The most likely reason for this is that the riddles were intended to be read aloud for others to solve; only the riddlemaster needed to see the answers. The following is a typical example of his art (Hovanec 1978: 14–15):

> *The Mule*
> *Unlike my mother, in semblance different*
> *from my father, of mingled race, a breed*
> *unfit for progeny, of others am I born, and none is born of me.*

So popular were Symphosius's riddles that they were plagiarized indiscriminately by other riddle-makers of the epoch. The exception was Metrodorus's *Greek Anthology* (c. A.D. 500), which contains original and truly challenging riddles, such as the following one that requires knowledge of Greek mythology (Hovanec 1978: 15):

> *If you put one hundred in the middle of a burning fire, you will find the son and a slayer of a virgin.*

(answer: *Pyrrhos*)

The riddle is solved by putting the Greek symbol for 100, *rho* [ρ], into the word *pyros* [πυρός], "fire." This produces the name *Pyrrhos* [Πυρρός], who was the son of Deidamia and the slayer of Polyxena.

A century and a half later, the English scholar and poet Aldhelm (640–709), inspired by Symphosius's style, produced a similar collection of a hundred riddles. It became one of the most widely quoted texts of the early medieval period. Aldhelm's riddles put on display the power of words to imbue everyday things (animals, plants, household items, etc.) with social meanings. The following riddle, for example, is a vivid metaphorical portrait of dogs (Hovanec 1978: 15):

> *The Dog*
> *Long since, the holy power that made all things*
> *So made me that my master's dangerous foes*
> *I scatter. Bearing weapons in my jaws,*
> *I soon decide fierce combats; yet I flee*
> *Before the lashings of a little child.*

Less well known are the riddles of Alcuin, the eighth-century scholar hired by Charlemagne to be his official puzzlist (chapter 1). Unlike Symphosius and Aldhelm, Alcuin did not give his riddles titles. The following one, which he sent to the Archbishop of Mainz, known by the nickname of Damoeta, is a typical example of Alcuin's riddling style (Hovanec 1978: 15):

> *A beast has sudden come to this my house,*
> *A beast of wonder, who two heads has got,*
> *And yet the beast has only one jaw-bone.*
> *Twice three times ten of horrid teeth it has.*
>
> *Its food grows on this body of mine,*
> *Not flesh, nor fruit. It eats not with its teeth,*
> *Drinks not. Its open mouth shows no decay.*
> *Tell me, Damoeta dear, what beast is this?*

(answer: *a comb*)

The solution to this riddle hinges on decoding the metaphorical meaning of "beast": a device rather than an animal. As this and the previous examples show, riddles may in fact be defined as extended metaphors, designed to provide figurative insights into the nature of things as perceived by their authors. This is perhaps why many poets have composed riddles. In Alcuin's era, for instance, the anonymous *Exeter Book* contained nearly a hundred examples of riddles that were composed as miniature poems about such everyday things as storms, ships, beer, books, and falcons.

Riddles also warn us of the power of metaphor to deceive. In the tenth century, a number of famous Arabic scholars used riddles specifically to alert people to the dangers that figurative language poses in written laws, coinciding with the establishment of the first law schools of Europe (Scott 1965). One such riddle-master was Al-Hariri (c. 1050–1120), whose collection (still untranslated into English, to the best of my knowledge) was called *Assemblies,* because it assembled in the same volume riddles and problems of grammar. Al-Hariri's riddles illustrate the ambiguities and hidden meanings that metaphor can introduce into legal discourse, and thus warn scholars against them.

In the fourteenth century, a Benedictine monk named Claret added a new, scandalous twist to the art of riddling. In a collection of 136 riddles that made him quite famous among the aristocrats of the era, he exploited double entendres, creating almost obscene riddles mischievously designed to induce lewd images in the mind of the solver. Here is a typical example of Claret's crafty riddling style (Hovanec 1978: 22):

> *A vessel have I*
> *That is round like a pear,*
> *Moist in the middle,*
> *Surrounded with hair;*
> *And often it happens*
> *That water flows there.*

(answer: *an eye*)

The image that the impish monk obviously wanted to elicit in the mind of his reader was that of the vulva, although the actual answer of an eye clearly fits the riddle's metaphorical description. By the late Renaissance, riddles were being tailored more and more to produce humorous or whimsical effects. One well-known English collection, *The Merry Book of Riddles,* was published in 1575. Here is an example from that work:

> *He went to the wood and caught it,*
> *He sate him downe and sought it;*
> *Because he could not finde it,*
> *Home with him he brought it.*

(answer: *a thorn in a foot*)

By the eighteenth century, riddles had become virtually every literate European person's favorite form of recreation, and were included as regular features of many newspapers and periodicals (Taylor 1951). Writers, poets, and philosophers constructed riddles to provide literary amusement. The American inventor Benjamin Franklin (1706–1790), too, composed many riddles under the pen name of Richard Saunders for inclusion in his *Poor Richard's Almanack,* first published in 1732. The riddles were a major factor in the almanac's unexpected success. In France, no less a literary figure than the great satirist Voltaire (1694–1778) would regularly compose mind-teasing riddles, such as the following one (Hovanec 1978: 28):

> *What of all things in the world is the longest, the shortest, the swiftest, the slowest, the most divisible and most extended, most regretted, most neglected, without which nothing can be done, and with which many do nothing, which destroys all that is little and ennobles all that is great?*

(answer: *time*)

Once again, metaphor is the key to unlocking the meaning of this riddle. Describing time as long, swift, slow, divisible, extended, etc. is so common in everyday discourse that we often forget that these expressions tell us virtually nothing about the true nature of time. They constitute metaphorical strategies for envisioning time as a physical entity.

The ever-increasing popularity and fashionableness of riddles at social gatherings brought about a demand for more variety. This led to the invention during the same century of a new riddle genre, known as the *charade* (the word may be derived from the Portuguese *charado,* "entertainment," the Italian *schiarare,* "to clear up," or the Provençal *charrado,* "chat or chatter"). The charade gradually gained popularity throughout Europe. Charades are solved one syllable or line at a time, by unraveling the double meanings suggested by the separate syllables, words, or lines, as the following example shows (Hovanec 1978: 61):

> *My first is to ramble;*
> *My next to retreat;*
> *My whole oft enrages*
> *In summer's fierce heat.*
> *Who am I?*

(answer: *a gadfly*)

The first line plays on the meaning of the word "gad," "to move about restlessly." The second line plays on the dual meanings of the verb "fly": "to flee or retreat from something" and "a dipterous insect." The last two lines of the charade complete the description of the gadfly as an irritating insect.

In the nineteenth century, this genre led to the *mime charade,* which became a highly popular game at social gatherings. It is played by members of separate teams who act out the syllables of a word, an entire word, or a phrase in pantomime. If the answer to the charade is, for example, "football," the syllables "foot" and "ball" are the ones normally pantomimed. A good description of the mime charade is found in William Thackeray's (1811–1863) great novel *Vanity Fair* (1848).

The charade made its way to America through the *Penny Post,* a magazine founded in Philadelphia in 1769. An American collection of riddles and charades based on familiar household objects, titled *The Little Puzzling Cap,* was published in 1787. By the nineteenth century riddles were firmly embedded in American recreational culture, having become regular features in newspapers, almanacs, and magazines of all kinds.

Given new life in both Europe and the United States in the nineteenth century were two ancient riddle genres known as *conundrums* and *enigmas.* Conundrums are riddles that exploit the similar sounds of word pairs such as "red" and "read," and the different meanings of words or expressions such as "all over" (Hovanec 1978: 28):

What is black and white and red all over?

(answer: *a newspaper*)

Enigmas (from the Greek "to darken and hide" or "to speak obscurely") are rhyming riddles that contain one or more veiled references to the answer, as the following famous enigma composed by the British statesman George Canning (1770–1827) clearly shows:

A word there is of plural number,
Foe to ease and tranquil slumber;
Any other word you take
And add an "s" will plural make,
But if you add an "s" to this,
So strange the metamorphosis;
Plural is plural now no more,
And sweet what bitter was before.

(answer: *cares—caress*)

The English politician and man of letters Horace Walpole (1717–1797) came up with the following truly ingenious enigma:

> *Before my birth I had a name,*
> *But soon as born I chang'd the same;*
> *And when I'm laid within the tomb,*
> *I shall my father's name assume.*
> *I change my name three days together*
> *Yet live but one in any weather.*

(answer: *today*)

Before its "birth," today does indeed have a different name—tomorrow. And when it is "laid within the tomb," i.e., when it is over, it takes a new name—yesterday. Finally, though it lasts only one day, it changes its name three days in a row ("three days together"): from tomorrow, to today, to yesterday.

Many of the ancient parables, the oracles of the Greeks, and the widely read fables of Aesop (c. 620–c. 560 B.C.) were composed in a largely enigmatic style. The great English author Jonathan Swift (1667–1745) was an ardent composer of enigmas, when he was not involved in his other literary pursuits. In the eighteenth century, riddles were also exploited for their educational potential. In that century *The Illustrated Book of Riddles* and various other riddle anthologies were published, primarily intended to entertain children during leisure hours. At the turn of the twentieth century, riddle-making in America had become a profitable enterprise in its own right, remaining so to this day.

As David Wells (1995: 169) points out, unlike mathematical puzzles, where twists or traps might hover fiendishly below the surface of the language in which they are framed, the appeal of riddles lies in the fact that the twist is built directly into the language itself, which, as we have seen, invariably denies words their literal meanings: "To the riddle *What is the difference between a hill and a pill?* the response *One is smaller than the other,* is not acceptable; the correct answer, recognizable by its twist, is *One is harder to get up, the other is hard to get down.*" The key to solving riddles lies, as Wells aptly remarks, in detecting the double entendre of the particular words with which they are framed.

To get a firmer grasp of what this means, it is useful to consider how a riddle can be created by exploiting the nonliteral meanings of a specific word. Take, for example, the word "smile." In English, a smile is commonly conceptualized metaphorically as something that, like clothing, can be worn. This is why we speak of wearing a smile, taking a smile off one's face, and so on. We can now use this metaphorical association between a smile and clothing to define a smile in the wording of a riddle as follows:

> *I am neither clothes nor shoes, yet I can be worn and taken off when not needed any longer. What am I?*

What makes this freshly invented riddle so irresistible is precisely the double entendre ruse that was employed to construct it. Its artfully contrived language is what creates "a web of antithetical relationships," as J. C. Heesterman (1997: 67) aptly puts it, in the mind of the solver.

The foregoing discussion raises the question of what function riddles have in human life. Are they just capricious forms of mental recreation? Or do they have some more profound message to convey? As many of the above examples demonstrate, riddles do indeed seem to harbor such a message. They warn us not to trust the greatest intellectual achievement of our species—language. The match between language and reality is hardly ever direct; rather, it is built on many intervening layers that are created by human cleverness. Solving riddles can be characterized as making one's way successfully through the gnarled webs of semantic layers in the language used to create them. No wonder, then, that in antiquity this ability was perceived as an indicator of intelligence. In the stories we tell children, too, riddles are commonly portrayed as tests of heroic cleverness. For example, in the anonymous *100 Riddles of the Fairy Bellaria*, published in 1892, a queen named Bellaria and her riddle-solving skills are pitted against a cruel invading king named Ruggero, who gives her a hundred riddles to solve. Ruggero threatens to destroy her empire should she fail to solve them successfully. The Riddler in the Batman stories is a modern-day descendant of Ruggero. Many stories composed for children are, in effect, either narrative riddles themselves, as are Aesop's fables, or else contain riddles, as do such widely known stories as Carroll's *Alice's Adventures in Wonderland* (1865) and *Through the Looking-Glass* (1871).

As the cagey medieval cleric Claret (page 41) knew, riddles can be composed on purpose to make fun of something or, simply, to provide a type of comic relief when we are in danger of taking life too seriously. Take, for example, the classic children's riddle "Why did the chicken cross the road?" The number of answers to this question is infinite. Among them are the following three:

(1) *To get to the other side.*
(2) *Because it was taken across by a farmer.*
(3) *Because a fox was chasing it.*

Answer (1) merely states the obvious, but might escape our attention at first because we have probably chosen to search for a nonobvious answer. Similarly, (2) and (3) are highly literal interpretations, but our mind seems inclined to resist them. All three answers are funny, and tend to provoke moderate laughter, similar to the kind that the punch line of some jokes elicits. This is because, as Wells (1988: 7) aptly points out, riddling and humor are both dependent "on a skeleton of double meanings, surprise, and the familiar in strange dress." Rather than an *Aha!* effect, such as many mathematical puzzles produce (chapter 1), riddles may thus be said to produce amusement, a *Ha, ha!* effect (see Paulos 1980, 1985).

Anagrams

Puzzle anagrams (Greek for "reversed letters") are of two types: (1) rearrangements of the letters of a word to make a new word ("tap" to "pat," "satin" to "stain," etc.); and (2) rearrangements of the letters in words to make a new phrase, expression, or sentence. Two widely cited anagrams of the second type by Lewis Carroll illustrate how they are constructed. One, using the letters of the name of the British humanitarian Florence Nightingale (1820–1910), provides her with a fitting eulogy; the other, using the letters of the name of the British political agitator William Ewart Gladstone (1809–1898), praises Gladstone's firebrand personality (cited by Costello 1996: 38):

(1) *Florence Nightingale* = *Flit on, cheering angel!*

(2) *William Ewart Gladstone* = *Wild agitator! Means well!*

One of the most famous anagrams of all time was constructed in the Middle Ages. Its unknown author contrived it as a Latin dialogue between Pontius Pilate and Jesus. Jesus' answer to Pilate's question "What is truth?" is phrased as an ingenious anagram of the letters of the words used by Pilate (Hovanec 1978: 67):

Pilate: *Quid est veritas?* ("What is truth?")
Jesus: *Est vir qui adest* ("It is the man before you.")

The origin of anagrams is shrouded in mystery. One thing is clear, however—in the ancient world, they were thought to contain hidden messages from the gods. Legend has it that even Alexander the Great

(356–323 B.C.) believed in their prophetic power. During the siege of the city of Tyre, Alexander was particularly troubled by a dream he had in which a satyr appeared to him. The next morning he summoned his soothsayers to interpret the dream. They pointed out to Alexander that the word "satyr" itself contained the answer, because in Greek "satyr" was an anagram of "Tyre is thine." Reassured, Alexander went on to conquer the city on the subsequent day.

Anagrams formed by taking the first letter of each line of a verse or group of words are called *acrostics* (from the Greek *akron*, "head," and *stikhos*, "row, line of verse") or sometimes *logogriphs*. The first four poems of the Book of Lamentations are acrostics. The first, second, and fourth, which are patterned after older funeral songs, comprise twenty-two verses, each beginning with a successive letter of the Hebrew alphabet; the third is an individual lament containing sixty-six verses, with three verses to each letter of the alphabet. The fifth poem is a group lament, again containing twenty-two verses, but it is not acrostic. All five laments speak movingly of God's harsh chastisement of his people for their sins. Coupled with these anguished verses are lines recalling God's mercifulness and expressing the prayerful hope that he may abate his wrath and restore a chastened Israel. The first part of the Book of Nahum (1:2–11) is also an unfinished acrostic poem, constructed with roughly half the letters of the Hebrew alphabet. The poem depicts God as jealous and angry, ready to take vengeance on those who oppose him. In the eighth section of the Book of Proverbs, a collection of short moral sayings, yet another twenty-two-line alphabetical acrostic poem can be found. And in Psalm 119, each eight-line stanza similarly begins with a successive letter of the Hebrew alphabet. Known appropriately as the Abecedarian Psalm, it is the oldest acrostic known.

The belief that anagrams were the linguistic vehicles used by the divinities to communicate with mortals was widespread. Many ancient prophets were essentially anagrammatists who interpreted this heavenly form of language. Soothsayer status was, in fact, often attained by those who claimed to possess knowledge of anagrams. In the third century B.C., for instance, the Greek poet and prophet Lycophron made a profession of devising anagrams of the names of the members of the Hellenistic king Ptolemy II's court in Egypt, as a basis for divining each person's character and destiny. For this, he became widely known and sought out as a soothsayer. Throughout the ancient world, it was commonly believed that personal names were anagrams foretelling the fates of individuals. People would often wear amulets with anagrams of their names on them to ward off evil.

Nowhere is the belief that letters hide fateful messages more evident than in the tradition of using *runes*—the ancient alphabetic characters of Germanic peoples—to predict the future. In elaborate rituals, tokens bearing the runes were put into a container and shaken. Placing one's hand into the container was said to produce an urge to select one rune in particular. The rune drawn was then thought to predict the individual's fate. For instance, the V symbol (*vara*) was a portent of healing (Vara was a Viking goddess who escorted the souls of warriors to Valhalla).

Anagrams have also been constructed retrospectively to explain a person's fate in life. For example, Mary Queen of Scots (1542–1587), who died by execution, was posthumously memorialized with the Latin expression *Trusavi regnis morte amara cada* ("Thrust by force from my kingdom I fall by a foul death"), which is an anagram of *Maria Steuarda Scotorum Regina* ("Mary Stuart Queen of Scots"). Shortly after Henry IV of France (1553–1610) was assassinated in 1610 by an unscrupulous man named Ravillac, it was pointed out, to everyone's amazement, that *Henricus IV Galliarum rex* ("Henry IV, King of the Gauls"), when rearranged, became *In herum exurgis Ravillac* ("From these Ravillac rises up"). Some writers preferred to use pen names made by rearranging the letters of their real names—perhaps believing in the power of anagrams to bestow fame and fortune upon them. The satirist François Rabelais (1493–1553), for instance, used the anagrammatic pen name Alcofribas Nasier, and Voltaire used Arouet le jeune, in which he substituted *V* for *U* and *I* for *E* (in Arouet): the first two words in this expression can be rearranged to form the name "Voltaire"—Arouet le (jeune) = Arovit + le = Voltaire.

Anagramming was transformed from an occult art to a form of intellectual recreation only after the Renaissance, when it became highly popular. In France, Louis XIII (1601–1643) even employed a Royal Anagrammatist to prepare challenging anagrams for him so that he could be intellectually stimulated during his leisure time. The post of Royal Anagrammatist came with a salary and a rank as high as any assigned to the famous poet laureates of the era. In the nineteenth century, Queen Victoria (1819–1901) popularized the *double acrostic,* making ingenious versions of her own, such as the one on the next page. The initial letters in the answer column, read downward, spell out the name of an English town, and the final letters, read upward, tell what that town is famous for:

A city in Italy	N	aple	S
A river in Germany	E	lb	E
A town in the U.S.	W	ashingto	N
A town in North America	C	incinnat	I
A town in Holland	A	msterda	M
The Turkish name of Constantinople	S	tambou	L
A town in Bothnia	T	orne	A
A city in Greece	L	epant	O
A circle on the globe	E	clipti	C

Lewis Carroll wrote scores of acrostic poems in the same century. The final poem in *Through the Looking-Glass* conceals the name of Alice. The first known anagrammatic book title, Samuel Butler's (1835–1902) *Erewhon*—which is made with the letters of "nowhere" nearly reversed—also comes from that century.

A *palindrome* (from the Greek for "running back again") is a particular type of anagram: a word or sentence that reads the same backward and forward. The reputed inventor of this genre was Sotades, a minor Alexandrian poet of the early third century B.C. of whom very little is known.

Word palindromes:
tot, mom, sis, ewe, non, madam

Sentence palindromes:
Madam, I'm Adam.
A man, a plan, a canal, Panama!

As Bombaugh (1961) has amply illustrated in his fascinating collection of word puzzles, palindromes are found throughout history and across languages. Many were apparently constructed on purpose for socially significant events.

A genre of anagram that is similar to the palindrome is the *word square,* a grid of seemingly random letters which, however, actually conceals words or messages. One of the earliest examples, dating from the second or third century A.D., is a 39 × 39 array of Greek letters carved in alabaster by an Egyptian sculptor known as Moschion. To read the square one must start at the center and read right or left, up or down, turning at right angles along the way. This reveals the phrase "Moschion to Osiris, for the treatment which cured his foot," which is repeated over and over in everlasting tribute to the healing god.

Another famous word square is the Sator Acrostic, found on the site

of the Roman city of Cirencester in England and on a column in the city
of Pompeii (Atkinson 1951):

```
R    O    T    A    S
O    P    E    R    A
T    E    N    E    T
A    R    E    P    O
S    A    T    O    R
```

Figure 2.1

The word *ROTAS* starts in the top leftmost cell and can be read across
the top row and vertically down the leftmost column; the same word
starts, in reverse order, in the bottom rightmost cell and can be read
from right to left across the bottom row and upward in the rightmost
column:

```
R---O---T---A---S
O    P    E    R    A
T    E    N    E    T
A    R    E    P    O
S---A---T---O---R
```

Figure 2.2

The word *OPERA* starts in the second cell of the top row and can be read
downward in the second column; it can also be found starting in the second cell from the top in the leftmost column, from where it can be read
from left to right across the second row. Moreover, it appears in reverse
order in the opposite cells: i.e., it starts in the second-to-last cell from the
bottom in the rightmost column, from where it can be read from right
to left; and it starts in the second-to-last cell from the right in the bottom row, from where it can be read upward:

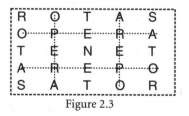

Figure 2.3

The remaining word in the square (TENET, third row, third column)
follows this mirror pattern. It is a palindrome. A generally accepted

translation of this acrostic is "Arepo, the sower, carefully guides the wheels," which is construed by some scholars as a metaphor for "God controls the universe" (Atkinson 1951; Bombaugh 1961: 348). This is why, throughout the early medieval period, it was commonly carved into amulets to ward off disease (Holroyd and Powell 1991: 78). But the word "Arepo" argues somewhat against this Christian interpretation because it appears to be of Celtic origin, meaning "plough." Moreover, since it has been found incised on a column in Pompeii, the only way to ascribe a Christian origin to the acrostic is to hypothesize that it was added to the column by Christians of a later period. Supporting the Christian interpretation of the acrostic is the fact that its letters were rearranged in the medieval period in the form of a cross to spell out *Pater Noster,* "Our Father." The two *A*s and *O*s on the arms of the cross stood for *alpha* and *omega,* the first and last letters of the Greek alphabet and thus, metaphorically, "the beginning" and "the end":

```
                        A

                        P
                        A
                        T
                        E
                        R
    A    P A T E R N O S T E R    O
                        O
                        S
                        T
                        E
                        R

                        O
```

Figure 2.4

The first word square in America appeared in *Wilkes' Spirit of the Times* in 1859. The same square appeared again in 1862 in *Godey's Lady's Book and Magazine.* Unlike the Sator Acrostic, it is not palindromic. The words it contains are "circle," "Icarus," "rarest," "create," "lustre," and "esteem" [see figure 2.5].

There are two modern-day descendants of the word square: (1) puzzles in which words are concealed within or across other words—a type of puzzle that, as far as I can tell, was probably invented by Lewis Carroll (Wakeling 1992: 27–29); and (2) puzzles in which words are hidden in a square array of letters. The sentences on the next page illustrate type (1). Each one contains a hidden female name (Hovanec 1978: 73) (see solution 2.1):

C	I	R	C	L	E
I	C	A	R	U	S
R	A	R	E	S	T
C	R	E	A	T	E
L	U	S	T	R	E
E	S	T	E	E	M

Figure 2.5

You can go by steamer to Canada.
This is the white camel I admire most.
He promises us a nice bottle of wine.
He saw them made at the palace.

The puzzle below exemplifies type (2), known more specifically as a *word-search* puzzle (see solution 2.2):

> *The names of five colors are concealed in the square below. They*
> *can be read from left to right, from right to left, upward and down-*
> *ward, and diagonally upward and downward.*

A	B	Y	E	L	L	O	W	N	M
H	J	K	L	M	N	O	P	N	B
V	M	N	N	E	E	R	G	G	L
B	N	M	N	K	O	P	L	K	U
S	W	E	I	B	N	M	K	Y	E
R	T	K	P	W	T	Y	U	M	N
Y	O	P	J	N	H	M	L	M	N
A	D	E	R	Q	U	I	I	E	O
N	M	E	L	P	O	M	T	B	V
X	X	X	R	A	S	D	C	E	X

Figure 2.6

Word-searches have become so popular today that entire magazines are devoted to them. They are also commonly used as pedagogical tools and games for teaching children words.

The question of why we continue to be fascinated by anagrams and their modern-day descendants demands some consideration, albeit brief, at this point. Anagrams are small-scale experiences—as puzzles were characterized generally in the previous chapter—of the use of symbols to understand the world. In a basic sense, every single word devised by human beings is an anagram for a fragment of knowledge, since it alludes to the presence of something real or imagined in the world. In my view, anagrams continue to appeal to the modern imagination because they are snapshots of how symbolism and knowledge are intertwined. As the great Romantic French poet Charles Baudelaire (1821–1867) aptly observed in his *Curiosités esthétiques* (1868: 3), "The whole visible universe is but a storehouse of images and signs to which the imagination will give a relative place and value; it is a sort of pasture which the imagination must digest and transform."

The deeply entrenched belief in the revelatory power of anagrams is curious, yet strangely compelling. Perhaps the ancients were right all along in claiming that anagrams were reifications of hidden messages, as we confront today one of the most difficult anagrams ever devised by Nature itself—the DNA. DNA, which is made up of sequences of chemical compounds called nucleotides, is the genetic material found within the cells of all living things. With the exception of identical siblings, each individual has a unique arrangement of nucleotides, and thus unique DNA. In a fundamental biological sense, therefore, the nucleotides that make up an individual's DNA are letters, anagrams of sorts that do indeed predict the individual's destiny.

Cryptograms

Cryptography is the art of concealing, encoding, or enciphering text. In his *History,* Herodotus (c. 484–425 B.C.) writes about a clever Greek spy named Demeratus who in 480 B.C. concealed a message under wax on wooden tablets so that it would not be detected by Persian guards. The message warned that Persia was planning a surprise attack on Greece. That story constitutes perhaps the oldest record of cryptographic writing in war. During the Peloponnesian War, about fifty years later, Spartan soldiers communicated with their field generals during battle by writing messages across a strip of parchment wrapped spirally around a staff called a *scytale.* When the parchment was unwound, the message became unreadable—except to the generals, who had another staff of the correct thickness and could rewrap the parchment around it. Cryptography as a science remained dormant, however, until the thirteenth century, when the great English scientist Roger Bacon (c. 1214–1292) described several coding systems and wrote a book completely in cipher, which, to the best of my knowledge, has never been deciphered. The first modern work on the science of cryptography was written by a German abbot named Johannes Trithemius in 1510.

Cryptography has played a central role in war throughout history. During World War I, for instance, British naval intelligence intercepted and deciphered a coded cable message, sent to Mexico in 1917 by the German foreign secretary Arthur Zimmermann, announcing German plans to begin unrestricted submarine warfare. That incident caused President Woodrow Wilson and Congress to declare war on Germany. During World War II, the British intelligence service hired thirty thousand people to break codes, including the British mathematician Alan Turing (1912–1954), a pioneer in computer theory. Because of security restrictions, Turing's role as a military cryptographer was not known until long after his death.

Today, cryptography has entered the computer age. Government agencies, banks, and corporations now routinely send a great deal of confidential information in cryptographic form from one computer to another. One of the first computer cipher systems, called LUCIFER, was devised in the early 1970s. In 1976, the National Institute of Standards and Technology developed a binary cryptographic technique (using fifty-six 0s and 1s to construct a key) called the Data Encryption Standard (DES). As there are more than 70 quadrillion possible keys, the chances of decoding a DES message would seem remote. This is why it was used in the first automated teller machines and in subscription television. Nevertheless, some experts in the field have challenged DES as being too vulnerable to very high power decoding methods developed for advanced computers. In 1978 the *RSA enciphering algorithm* was developed, based on two 100-digit prime numbers.

Given the extraordinary feats of ingenuity that have gone into inventing and breaking secret codes in wartime and the protection of information, it is little wonder that the appeal of cryptography has extended beyond the military and bureaucratic realms, motivating a genre of puzzles called, appropriately enough, *cryptograms* (Gaines 1956; Kahn 1967; Gardner 1972; Singh 1999). In such puzzles, the original message is called the *plaintext,* its encoded form the *ciphertext.* The method that changes one into the other is termed the *cryptosystem.* In order to convert the ciphertext back into plaintext, the solver has to figure out the cryptosystem. *Encrypting* is the term used to refer to the process of converting plaintext into ciphertext, and *decrypting* refers to the opposite process of changing ciphertext into plaintext.

The three most common encrypting methods used by puzzlists to construct cryptograms are (1) replacing the letters in a message with numbers (a *letter-to-number cryptogram*); (2) replacing the letters with other letters (a *letter-to-letter cryptogram*); and (3) replacing the letters with specific types of symbols (a *letter-to-symbol cryptogram*). The following is an example of the first type, also known as a *Polybius cipher* after the Greek historian Polybius (c. 200–c. 118 B.C.), who was among the first to encrypt messages by converting letters to numbers. It is discussed in detail here because its solution illustrates the reasoning process involved in solving all types of cryptograms (Danesi 1997: 129):

> *Decipher the following message, given that* 2 = S, 6 = P, 10 = B, *and* 12 = K. *The message relates to a type of reading material.*

> *1 0 4 2   4 2   5   6 3 7 7 8 9   10 11 11 12*

We can start by replacing the numbers 2, 6, 10, and 12 with the given letters. This provides an initial view of the hidden message:

> *1 0 4 S   4 S   5   P 3 7 7 8 9   B 11 11 K.*

Since 5 is a single-letter word, we conclude that it can only be the pronoun *I* or the indefinite article *a*, because no other single-letter English words exist. It is best to start out with the most likely substitution, which in this case is 5 = A. The pronoun *I* usually starts a sentence: "I am going out"; "I love puzzles"; etc. There are, of course, sentences where this is not the case. But they are less frequent. In any case, we can always go back to this point in our solution if our initial replacement should not work out, and then try again with 5 = *I*. Our evolving message now looks like this:

> *1 0 4 S   4 S   A   P 3 7 7 8 9   B 11 11 K.*

The last partially deciphered word in the sentence, *B 11 11 K*, contains a sequence of two identical letters, represented by two *11*s. These can only be vowels, because English words must contain vowels and we know that the other two letters are consonants.

There are, of course, five substitution possibilities for the two *11*s: *aa, ee, ii, oo,* or *uu*. The only replacement that produces a legitimate word, however, is *oo*. The hidden word is, therefore, *book*. Our emerging plaintext now looks like this:

> *1 0 4 S   4 S   A   P 3 7 7 8 9   B O O K.*

The 4 in the two-letter word *4 S*, by the same logic, can only be a vowel. So the word could be *as, is,* or *us*. The partially revealed plaintext suggests that *is*, being a verb, is the most likely possibility. Once again, should this turn out to be erroneous, we can always go back to this point in our solution and try out *as* or *us* to see where they would lead. For now, we replace *4* with *I* both here and in the first word as well:

>   *1 0 I S   I S   A   P 3 7 7 8 9   B O O K.*

The only meaningful substitution for *1 0 I S*—in combination with the second word *is*—is the demonstrative pronoun *this:*

>   *T H I S   I S   A   P 3 7 7 8 9   B O O K.*

Only the word *P 3 7 7 8 9* remains to be deciphered. Here is what we now know about it: (1) It refers to a kind of book. (2) It is spelled with six letters, starting with *P*. (3) It has two identical letters in it. Now, using the clue given by the puzzle—*The message relates to a type of reading material*—we can easily conclude that *puzzle* is the required word. So the plaintext is: *This is a puzzle book.*

As this puzzle shows, the successful solution to a Polybius cipher hinges on knowing linguistic facts—e.g., what single-letter words are possible in a language, what sequences of double letters, what final letters, and so on. The same type of knowledge is required to solve letter-to-letter cryptograms, called *Caesar ciphers* (because Julius Caesar supposedly invented them). In the following puzzle, for instance, the key linguistic insight is that the letters in the plaintext have been shifted two spaces later in the normal alphabetic sequence (Hovanec 1978: 58):

>   *J G N N Q*

That is, *J* was used for *H*, *G* for *E*, *N* for *L*, and *Q* for *O*. Making the appropriate substitutions produces the answer:

>   *J G N N Q = H E L L O*

Ingenious techniques for constructing Caesar ciphers are found scattered throughout the annals of history. The sacred Jewish writers of ancient times, for instance, concealed their messages by reversing the alphabet, i.e., by using the last letter of the alphabet in place of the first, the next-to-last in place of the second, and so on. This cryptosystem, called *atbash*, is exemplified in Jeremiah 25: 26, where one finds the word "Babel" (Babylon) encrypted as "Sheshech" (Beck 1997). The American president Thomas Jefferson (1743–1826) built an ingenious device for making Caesar ciphers, consisting of thirty-six wooden wheels, each representing the letters of the alphabet printed in different arrangements. A message could be enciphered automatically by lining it up against the wheels.

The most challenging Caesar ciphers are, needless to say, random substitutions, as the following puzzle illustrates (Hovanec 1978: 62):

> *DFC KAIUR KYUU XCMCI WDHIMC TAI KAXRCIW, NJD*
> *AXUQ TAI KHXD AT KAXRCI*

The key to solving this cryptogram is to determine where the articles are, which two- and three-letter words ("at," "to," "the," "our," "for," etc.) are likely to be used in specific positions, etc., keeping always in mind various facts about the language in question. For example, in English the most common letter at the end of a word is *e*, *q* is always followed by *u*, double letters between consonants must be vowels, etc. Noting that the first word has three letters, we can start by assuming that it stands for the definite article "the." As above, should this turn out to be erroneous, we can always go back and try out different possibilities. This allows us to establish that $D = T$, $F = H$, and $C = E$. The substitutions are underlined:

> *THE KAIUR KYUU XEMEI WTHIME TAI KAXREIW, NJT*
> *AXUQ TAI KHXT AT KAXREI*

Since the cipher word *AT* cannot stand for itself, the plaintext "at," it must represent another two-letter word. Considering simultaneously the *A* and *T* in the form *TAI* suggests $AT = OF$ as the most likely substitution $(A = O, T = F)$, and thus that $I = R$ (since "for" is a very common English word):

> *THE KORUR KYUU XEMER WTHRME FOR KOXRERW,*
> *NJT OXUQ FOR KHXT OF KOXRER*

The form *KYUU* appears in a position which suggests that it is probably a verb form. Since it has a double *U* at the end, it probably stands for the modal verb "will." Again, should this turn out to be incorrect, we can always go back to this point in the puzzle and try out a different hypothesis. We can now make the appropriate replacements $(K = W, Y = I, \text{ and } U = L)$:

> *THE WORLR WILL XEMER WTHRME FOR WOXRERW, NJT*
> *OXLQ FOR WHXT OF WOXRER*

It is obvious that the form *WORLR* stands for *WORLD*, and thus that $R = D$. The form *XEMER* can only be deciphered as *NEVER* $(X = N$ and $M = V)$. Readers can confirm this for themselves by trying out different substitutions:

THE WORLD WILL NEVER W TH RVE FOR WONDERW, NJT
ONLQ FOR WH NT OF WONDER

The rest of the cipher can be easily decoded. The plaintext is *The world will never starve for wonders, but only for want of wonder.*

Solving letter-to-symbol ciphers involves the same reasoning process as solving Polybius or Caesar ciphers. Take, for example, the following simple cryptogram, whose solution is left as an exercise for the reader (see solution 2.3):

    #   A $ # * A L # A N

Letter-to-symbol ciphers were used in the Middle Ages by alchemists, who employed astrological signs to encipher their secret messages. But the most famous of all methods of encipherment was devised not by some occult scholar, but rather by the American inventor Samuel Finley Breese Morse (1791–1872). Known as the *Morse Code,* it is a system of radiotelegraphic signals, which was used widely in the nineteenth and early twentieth centuries. The code consists of combinations of dots and dashes, or short and long tones, representing the letters of the alphabet and digits. Today this code is rarely used, because radiotelegraphy has been replaced by printing telegraph systems, facsimile transmission, and other new forms of electronic communication, especially the Internet.

A widely used method of enciphering messages, known as *transposition,* involves transposing or reordering the plaintext letters in some way. As an example, consider the following plaintext (from Pappas 1991: 75):

MEET MARTHA ON MONDAY IN FRONT OF THE BRIDGE

This can be enciphered in several ways: (1) by writing the words backward (*TEEM AHTRAM NO YADNOM NI TNORF FO EHT EGDIRB*); (2) by separating the vowels from the consonants (*MTEE MRTHAA NO MNDYOA NI FRNTO FO THE BRDGIE*); (3) by reversing syllables wherever possible (*MEET THAMAR ON DAYMON IN FRONT OF THE BRIDGE*); and by other transformations. Messages can also be enciphered by a combination of substitution and transposition methods. For example, the letters of the above message can be assigned number values in the order they appear ($M = 1$, $E = 2$, $T = 3$, $A = 4$, $R = 5$, $H = 6$, $O = 7$, $N = 8$, $D = 9$, $Y = 10$, $I = 11$, $F = 12$, $B = 13$, $G = 14$), and this substitution method can then be applied to every odd word (the first, third, fifth, seventh, and ninth). At the same time, the transposition method of writing the words backward can be applied to every even word (the second, fourth, sixth, and eighth). This would produce the following ciphertext:

> *1 2 2 3   AHTRAM   7 8   YADNOM   11 8   TNORF   7 12*
> *EHT   13 5 11 9 14 2.*

As the reader can now see, it would be a difficult puzzle indeed to solve, involving a solid knowledge of language structure together with quite a bit of insight thinking. Decrypting this puzzle imparts a feeling akin to that of solving a mystery. This is perhaps why many great mystery writers have used cryptograms in their plots as elements of suspense (Rosenheim 1997). In *The Gold Bug,* for example, Edgar Allan Poe included a random-substitution letter-to-symbol cipher, using numbers and punctuation marks, and written without word divisions, supposedly devised by the pirate Captain Kidd. In his Sherlock Holmes mysteries, Sir Arthur Conan Doyle (1859–1930) frequently included ciphers for the master detective to unravel. The 1930s fictional crime fighter The Shadow used special symbols to make his own ciphers, adding considerably to the aura of mystery that surrounded his character. The great writers of detective fiction have always known that the intricate process of decoding a cipher—one letter-clue at a time—adds immeasurably to the suspense of the story.

In a fundamental sense, the methods of all the modern sciences are akin to decryption, since unlocking the mysteries of Nature is strikingly similar to solving a cryptogram. The famous German philosopher and mathematician Gottfried Wilhelm Leibniz (1646–1716) in fact characterized scientists as cryptographers, suggesting that their primary objective was to crack the infinite array of ciphers that Nature presented to them. Nature is a large-scale code. It can only be decoded by small-scale procedures, as we have called puzzles in this book, that scientists call *theories*. Theories are essentially special kinds of cryptosystems devised by scientists to help them unlock the innumerable puzzles that Nature presents to them.

Rebuses

One type of cipher, known as a *rebus*, requires separate treatment. A rebus is a message enciphered by replacing words or parts of words with pictures, signs, letters, etc. The following is a rebus of the word "island"— 👁, an eye, represents the vowel *i* in "island"; the ampersand sign *&* represents the word "and"; and when the pieces are joined together, the word "island" results:

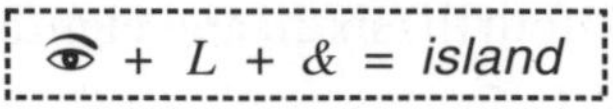

$$👁 + L + \& = island$$

Figure 2.7

Rebuses have been a source of fascination throughout the world and across history. It is not known where, when, or why they originated. Coins with rebuses inscribed in them, representing famous people or cities, were common in ancient Greece and Rome (Céard 1986). During the Middle Ages, rebuses were frequently used to encode heraldic mottoes. In Renaissance Italy, Pope Paul III (1468–1549) employed rebuses to teach writing. In the early part of the seventeenth century, the priests of the Picardy region of France put them on the pamphlets they printed for the Easter carnival, so that even the illiterate masses could understand parts of the message. So popular had rebuses become throughout Europe that Ben Jonson (1572–1637), the English playwright and poet, trenchantly ridiculed them in his play *The Alchemist*. Rebus cards appeared for the first time in 1789 (Costello 1996: 8). By the nineteenth century, they had become highly popular. Today, rebuses are found everywhere—in advertising, in logos, in puzzle collections, etc. Like riddles, they have always been especially appealing to children. The following two, in fact, are taken from an 1865 issue of a popular nineteenth-century children's magazine titled *Our Young Folks* (Hovanec 1978: 68). They refer to two English writers: the first one to Alfred, Lord Tennyson (1809–1892) and the second to Thomas Carlyle (1795–1881). Their solutions are indicated below (note that the *X* is to be read as the Roman numeral ten):

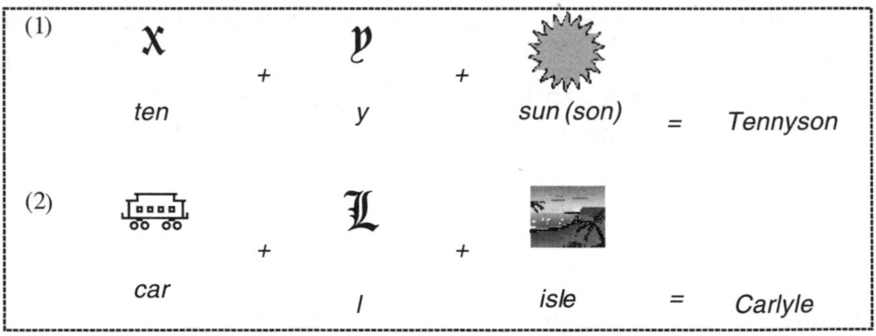

(1) X	+	y	+	sun (son)	=	Tennyson
ten		y				
(2)	+	L	+	isle	=	Carlyle
car		l				

Figure 2.8

Some rebuses rely on the physical layout or relative positions of certain symbols or words on the page, as the following two examples show (Hovanec 1978: 41):

(1)

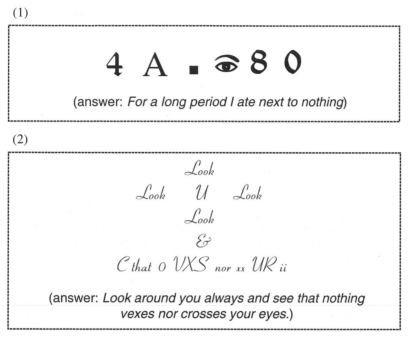

(2)

(answer: *Look around you always and see that nothing vexes nor crosses your eyes.*)

Figure 2.9

A rebus of this type was reportedly sent by Frederick the Great (1712–1786), King of Prussia, to Voltaire (Costello 1996: 8):

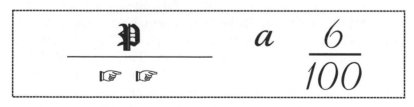

Figure 2.10

This translates in French to *Deux mains* ("two hands") *sous* ("below") "*p*" (the name of this letter is pronounced "pay" in French) *à* ("at") *cent* ("one hundred") *sous six* ("below six"): i.e., *Demain souper à Sans Souci* ("Dinner tomorrow at Sans Souci"). Voltaire answered the invitation with his own rebus:

$$\mathfrak{G} \ \text{a grand a pétit}$$

(answer: *J'ai grand appétit* "I have a big appetite")

Figure 2.11

Combining pictographic with alphabetic symbolism, rebuses put on display the power of human signs to encode meaning. A pictographic representation may stand for its referent, the thing it depicts (as the zero in Figure 2.9 stands for the concept "nothing"), or it may stand for the pronunciation of its referent's name (as the 👁 stands for the sound *ah-ee* and thus the pronoun "I"). An alphabetic representation may similarly stand for either its referent, its meaning as a word or letter (as the words "look," "that," and "nor" in Figure 2.9, and the letters *y* and *l* in Figure 2.8, stand for themselves), or its pronunciation (as "a pétit" in Figure 2.11 is meant to be sounded out as the word "appétit"). But alphabetic characters still have an embedded pictographic element in them—an element that rebuses bring out concretely. Rebuses thus remind us that we encode a large part of our knowledge about the world through visual symbols, not just written words.

Crosswords

Of all the puzzles discussed so far in this chapter, the *crossword* is the only one that does not have ancient origins. It is a twentieth-century invention, devised not for some mystical or occult reason but for the sole purpose of providing intellectual entertainment. It was created by Arthur Wynne, who was born in Liverpool, England, in 1871 and immigrated to the United States in 1905. As editor of the "Fun" section of the *New York World,* a Sunday color supplement, Wynne introduced what he called a *Word Cross* on December 21, 1913, after he had seen something similar to it in England. Because of a compositor's error, the title became *Cross Word* in its third week, and the name has stuck ever since. Here is Wynne's original crossword (the word "fun" was given as a "free word"). The words are to be guessed from the given clues and fitted into the interlocking grid of horizontal and vertical squares (see solution 2.4). [See figure 2.12.]

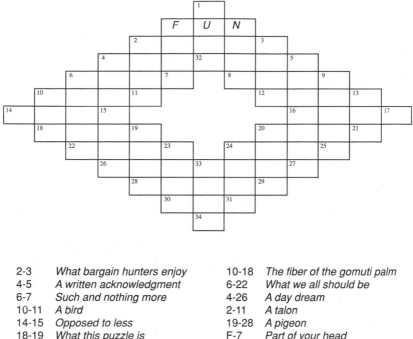

2-3	What bargain hunters enjoy		10-18	The fiber of the gomuti palm
4-5	A written acknowledgment		6-22	What we all should be
6-7	Such and nothing more		4-26	A day dream
10-11	A bird		2-11	A talon
14-15	Opposed to less		19-28	A pigeon
18-19	What this puzzle is		F-7	Part of your head
22-23	An animal of prey		23-30	A river in Russia
26-27	The close of a day		1-32	To govern
28-29	To elude		33-34	An aromatic plant
30-31	The plural of is		N-8	A fist
8-9	To cultivate		24-31	To agree with
12-13	A bar of wood or iron		3-12	Part of a ship
16-17	What artists learn to do		20-29	One
20-21	Fastened		5-27	Exchanging
24-25	Found on the seashore		9-25	To sink in mud
			13-21	A boy

Figure 2.12

Unexpectedly, readers inundated Wynne with requests for more puzzles of this type. Overnight, the crossword became a craze in the city of New York. By 1924, crosswords had grown into a national pastime in the United States, gradually taking on the more familiar appearance of patterned black and white boxes, with two sets of clues, one for words to be written horizontally (*across clues*) and the other for words to be written vertically (*down clues*). That year, the American publishing company Simon and Schuster printed the first book compilations of crossword puzzles; each book came equipped with a pencil and eraser and a penny postcard, which buyers could mail to the publisher to request the answers. The publisher claimed to have sold 400,000 copies of the first three volumes during 1924. To take advantage of the crossword mania, manufacturers soon began making jewelry, dresses, ties, etc. with crossword designs on them. A song called "Crossword Mama, You Puzzle Me, but Papa's Gonna Figure You Out" was the B side of a popular 1924 record. A Broadway play, *Puzzles of 1925*, which dealt with the crossword puzzle craze, met with resounding success. The heart of the play featured a scene in a "Crossword Puzzlers Sanitarium" for people driven insane by their obsession with crossword puzzles. Shortly thereafter, board games were invented incorporating the crossword puzzle idea. Of these, Scrabble has remained the most popular.

But despite its apparent novelty, the crossword was really no more than a twentieth-century descendant of the acrostic puzzle. Its immediate progenitor was, in fact, a popular British late-nineteenth-century acrostic, designed specifically for children. In 1925, at the height of the crossword craze, Wynne acknowledged as much in the following words (quoted in Costello 1996: 22):

> I awakened recently to find myself acclaimed as the originator of the crossword puzzle, which everybody is doing now. But all I did was take an old idea as old as language and modernize it by the introduction of black squares. I'm glad to have had a hand in it, and no one is more surprised at its amazing popularity.

Another nineteenth-century forerunner of the crossword was Lewis Carroll's *doublet* puzzle, also called a *word ladder* (Knuth 1999). This puzzle presents two words to the solver, who is required to change one to the other by changing only one letter at a time, forming a legitimate new word with each change (Salny and Frumkes 1986: 82–83). Carroll published many doublet puzzles in *Vanity Fair* between 1879 and 1881 (Wakeling 1992: 39–41). The following is an example of Carroll's invention:

> *Turn the word HEAD into TAIL, by changing only one letter at a time, forming a new word each time to do so.*

Here is one solution:

Head
↓
Heal
↓
Teal
↓
Tell
↓
Tall
↓
Tail

The magazine, incidentally, offered weekly prizes for solutions. Carroll later amended the rules somewhat to add variety and, thus, stimulate more interest in his ludic invention, as can be seen in the following puzzle he created in 1892 (see solution 2.5):

> *Change IRON into LEAD by introducing a new letter or by rearranging the letters of the word, at any step, instead of introducing a new letter. But you may not do both in the same step.*

Word puzzles similar to the crossword, in the form of squares, diamonds, pyramids, and other figures, had long before been presented with clues. Wynne's innovation was to include a grid of squares to be filled in. The level of difficulty rose quickly, as skilled crossword solvers started demanding much more challenging clues. Special dictionaries soon had to be compiled of unusual words that had found their way into crossword clues.

The three individuals who developed the crossword puzzle into its contemporary forms were Margaret Petherbridge Farrar, Edward Powys Mather, and Elizabeth S. Kingsley (Millington 1974). Margaret Petherbridge (1897–1984) became puzzle editor of the *New York World* in 1921, where she produced ingenious, challenging crosswords and garnered a large readership. She was also the coeditor (with Prosper Buranelli and F. Gregory Hartswick) of the first few Simon and Schuster crossword puzzle books. She continued to edit the Simon and Schuster books until her death, by which time the series had reached volume 136. The royalties, incidentally, allowed her to loan the editor John Farrar, whom she had by now married, sufficient capital to start two publishing houses, Farrar and Rinehart in 1929 and Farrar, Straus, and Company in 1946 (which later became Farrar, Straus & Giroux). She became the first crossword puzzle editor at the *New York Times* in 1942 and held the post until 1969; in the same period she also created crosswords for such popular magazines as *Esquire, Seventeen, Sports Illustrated,* and *Good Housekeeping.* She was responsible for standardizing the symmetrical grid format that most crossword puzzles have today, with black and white boxes and across and down clues. She also introduced the now common practice of including quotations, puns, and all kinds of word play in clues.

The British crossword-maker Edward Powys Mather, known to his puzzle audience as Torquemada (the name of a notorious Spanish Grand Inquisitor), introduced the technique of concealing the answer to a clue within the clue:

> *Material used in many long dresses*
>
> (answer: *nylon*)

Mather composed crossword puzzles for the *Saturday Westminster* in 1925 and then for the *Observer* from 1926 to 1939. He soon gained a reputation as one of the toughest crossword puzzlists of all time. Here are a few examples of his formidably challenging clues:

> *The artist has been about cooked with herbs*
>
> (answer: *saged*)
>
> *Marriage days*
>
> (answer: *weds*)
>
> *Tell the physician to love, and then he's quite often my subject*
>
> (answer: *drama*)
>
> *We can't honestly say there's no harm in it, but we're glad the French girl's got it*
>
> (answer: *charm*)

Elizabeth S. Kingsley (1871–1957) introduced the *double-crostic* in 1934 in the *Saturday Review of Literature.* This version of the crossword genre involves guessing words defined in the puzzle and then writing them in the given squares of a diagram, in which a quotation, the name of a famous personage, the title of some work, etc. comes into view when the correct letters are supplied. Here is a simple example of this type of puzzle (see solution 2.6):

By answering the clues correctly, you will be able to read the name of a famous composer, from top to bottom, in the first column. The clues relate to the composer.

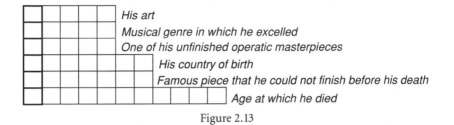

His art
Musical genre in which he excelled
One of his unfinished operatic masterpieces
His country of birth
Famous piece that he could not finish before his death
Age at which he died

Figure 2.13

Today, the crossword is probably the most popular of all puzzles. More magazines and books are published each year for crossword puzzlists than for fans of any other puzzle genre. The puzzle instinct's penchant for pattern and symmetry is something that the crossword seems to indulge quite nicely. The grid, with its symmetrical rectangular configuration crying out for missing letters, has hooked millions of humans across the globe, who obviously cannot tolerate leaving it in a state of emptiness. Many stories are told, in fact, of crossword puzzle addiction. According to one story, two New York magistrates had to ration crossword addicts who had neglected their families to two puzzles a day (Costello 1996: 23). Another tells of a human life put in jeopardy because of crossword-puzzle deprivation (Hovanec 1978: 109):

> During one especially debilitating snowstorm when service to rural areas was completely cut off, a nurse at a hospital in one of the affected areas wrote to Mrs. Farrar and told her that a patient's recovery was in jeopardy unless she received a copy of the answers to a previous puzzle. It seemed that this patient was recuperating through solving the crossword found in the Times.

Why is it, one may ask, that such a trifling thing has such an overpowering influence on human beings? There is, of course, no simple answer to this question. All that can be said is that the puzzle instinct continues to have a stranglehold on the imagination. The crossword puzzle mania is, in effect, a product of an instinctual need to search out definite, reassuring, small-scale answers to the many questions that life presents. Filling in the little squares with clear-cut answers seems, in its own minuscule way, to negate the existential emptiness that human beings unconsciously feel.

The Janus-Faced Nature of Language

Lewis Carroll was fascinated by the boundless range of meanings that language has the potential to encode. By inventing new words, he wanted to show his young readers that words fill in the many gaps that we otherwise would feel exist in the language. For example, he coined "brillig," "slithy," "toves," and "wabe" to mean, respectively, "the time of broiling dinner at the close of the afternoon," "lithe and slimy," "animals that are something like badgers—they're something like lizards—and they're something like corkscrews," and the "grass plot around a sundial." These were concepts for which no specific words existed. Carroll demonstrated, in effect, that the structure we use to make words provides us with the means to play an endless anagrammatic game with reality. The result of such play is meaning. As two great twentieth-century linguists, Edward Sapir (1884–1939) and Benjamin Lee Whorf (1897–1941), showed in their ground-breaking research (Sapir 1921; Whorf 1956), human ideas are, as Carroll suspected, rooted in the structure of language. And indeed, if English-speaking people started using the words "brillig," "slithy," "tove," and "wabe" routinely, then after a while they would start seeing or recognizing brilligs, slithy things, toves, and wabes everywhere, eventually believing that the things they refer to must exist. For that matter, Carroll's word "chortle," which he coined by combining "chuckle" and "snort," has passed into common usage, and we now chortle and hear others do so.

All this in no way implies that language constrains or stifles the imagination. On the contrary, as we have witnessed in this chapter, the riddles, anagrams, acrostics, cryptograms, and other linguistic puzzle forms that the puzzle instinct has engendered in this domain of Puzzleland make it obvious that language is a malleable instrument that can be put to any use the human imagination desires. Should the need arise to create a new word-category, all we have to do, as Carroll showed, is be consistent with the structural requirements of our language's sound system.

The presence of word puzzles in human life puts on display the "Janus-faced" nature of language, which, as Sapir put it, is paradoxically both limiting and limitless. Janus was a Roman god who, as guardian of gates and doorways, had two heads that allowed him to look in different directions at the same time. Fascinated by the Janus-faced nature of words, an association called Oulipo—an acronym for "Ouvroir de littérature potentielle" (Workshop of potential literature)—was founded in Paris in the mid-twentieth century by a small group of writers and mathematicians devoted to exploring the relation between language and reality (Paulos 1991: 166–168). Their activities were (and continue to be) truly intriguing, and highly relevant to the subject matter of this chapter. For example, Raymond Queneau, one of the founders of Oulipo, published a book of poetry titled *100 Trillion Sonnets,* consisting of ten sonnets, one on each of ten pages. The pages are cut so as to allow each of the fourteen lines of a sonnet to be turned separately. The physical format of the book allows 100 trillion combinations of lines—100 trillion sonnets. All of them, Queneau claimed, "make sense."

Another example of an oulipian work is Georges Perec's three-hundred-page novel *La disparition.* No word in the novel contains the letter *e.* Such a work, omitting a letter or letters, is called a *lipogram.* Perec insisted that his lipogrammatic novel was worthy of being considered literature, since it was designed to explore language's infinite possibilities. Incidentally, the celebrated American humorist James Thurber (1894–1961) also wrote a lipogrammatic work in 1957, titled *The Wonderful O.* It was a political fable for children, telling what happened when Captain Black, a literate pirate who hated the letter *o,* banished the letter from the island of Ooroo.

Another oulipian technique is to make up sentences, called *pangrams,* which contain all the letters of the alphabet. Here are a few examples of pangrams (Bombaugh 1961: 326):

> *Pack my box with five dozen liquor jugs* (32 letters)
> *A quick brown fox jumps over the lazy dog* (33 letters)
> *Waltz, nymph, for quick jugs vex Bud* (28 letters)
> *Quick wafting zephyrs vex bold Jim* (29 letters)

Lipograms and pangrams are artifacts showing that a constraint on linguistic structure in no way constrains the making of meaning. The oulipian approach to linguistic creation shows us that we are inclined to extract meaning from words even when we artificially restrict the ways they can be made (without certain letters, using all the letters of the alphabet, etc.). The ancient anagrammatists sought to uncover hidden meanings in virtually the same way—by playing with word forms (rearranging their letters). Incidentally, oulipian-type experiments seem to have been of interest to the ancients as well. Around 2500 years ago, the Greek poet Pindar (c. 522–c. 443 B.C.) created puzzle poems that followed a certain rule or bore a hidden message (Costello 1996: 5). For example, he wrote an ode without using the letter *sigma* (*s*). Such experiments with language hold a caveat for us, as do riddles and the other kinds of puzzles discussed in this chapter. They warn us that language is Janus-faced, a versatile instrument for encoding reality as well as a weapon that can be used to obfuscate the same reality.

3 Puzzling Pictures

OPTICAL ILLUSIONS, MAZES, AND OTHER VISUAL MIND-BOGGLERS

Mathematics would certainly have not come into existence if one had known from the beginning that there was in nature no exactly straight line, no actual circle, no absolute magnitude.
—Friedrich Nietzsche (1844–1900)

To the naked eye, the opposite angles formed when two straight lines intersect appear to be equal. But this appearance, as obvious as it may seem, does not establish the angles as necessarily or always equal. In his great textbook of mathematical theory and method, known as *Elements*, the Greek mathematician Euclid (who lived around 300 B.C.) went beyond the evidence of eyesight and proved, beyond any iota of doubt, the equality of such angles. Using an argument that is remarkably easy to follow, Euclid demonstrated that the opposite angles formed when two straight lines intersect not only look equal to the eye, they are necessarily so.

On the next page is a diagram similar to the one used by Euclid in his demonstration. The intersecting lines are labeled AB and CD, and two of the four opposite angles formed by their intersection are labeled x and y. The angle between x and y on the top part of the diagram is labeled z. The goal is to show that x and y, no matter what their size, will always be equal [see figure 3.1].

Euclid's proof can be given in contemporary algebraic notation as follows. He started by reminding his readers that a straight line is in fact an angle of 180°. Since CD is composed of two smaller angles, x and z,

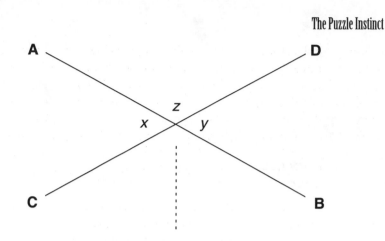

Point of Intersection

Figure 3.1

the sum of the two is 180°—a statement that in modern algebraic notation can be represented with the equation $x + z = 180°$. Also, since AB is
composed of y and z, their sum, too, is 180°—a statement that can be
represented with the equation $y + z = 180°$:

(1) $x + z = 180°$
(2) $y + z = 180°$

Equations (1) and (2) can be reformulated as follows:

(3) $x = 180° - z$
(4) $y = 180° - z$

Logically, since things equal to the same thing are equal to each
other, we can deduce that $x = y$, since both variables are equal to the
same thing, namely $(180° - z)$. Euclid ended his proof with the declara-
tion "Which was to be demonstrated," a phrase expressed later with the
Latin abbreviation *QED*, for *Quod erat demonstrandum*—the stamp of
mathematical authority that Euclid's method of demonstration came
to represent. To the emerging mathematical way of thinking, Euclid's
QED method was incontestable, because it demonstrated by logical argument why a certain pattern perceived as noteworthy and regular by the
eye is the way it is.

Although the early builders and land surveyors measured fields and
laid out right angles with strings and various instruments, they relied
primarily on visual inference. Nonetheless, using the patterns that they
perceived with their eyes, they were able to construct truly impressive

buildings and accomplish remarkable engineering feats. Those ancient architects and engineers suspected, of course, that many of the patterns they saw had an underlying logic. But unraveling such logic had to await the groundbreaking approach of Euclid, who systematized the attempts of mathematicians such as Pythagoras and Thales (c. 662–c. 545 B.C.) before him to reconcile sensory perception with rational understanding. As Kline (1959: 75) aptly puts it, "Euclid was the great master who arranged the scattered conclusions of his predecessors so that they all followed by deduction."

Needless to say, the relation between visual perception and accurate reckoning, to use Ahmes' term, was not the exclusive jurisdiction of the ancient Greek geometers. As a matter of fact, it has always been of great interest to mathematicians and puzzlists alike, all over the world. The puzzles that the latter have left for posterity constitute playful experimentations with visual pattern and the QED method. On this leg of our journey through Puzzleland, we thus come to the region inhabited by the Carrollian character of the Caterpillar, where things, like the insect itself, are not always what they seem to be. Here puzzlists, with their many ingenious tricks and traps in hand, have been busily working away since antiquity to expose eyesight as sometimes unreliable—the very same objective that Euclid and other great geometers had when they established geometry as a deductive science.

Visual Trickery

We start our journey by considering the innumerable visual tricks created by both artists and puzzlists to deceive the eye. Not only are these interesting in themselves, they also hold the same kind of warning about visual perception that riddles hold about verbal perception (chapter 2) —namely, that things are not always what they appear to be.

Visual tricks can be divided into two main categories: *picture incongruities* and *optical illusions*. The former can be further subdivided into (1) pictures that conceal disguised images; (2) pictures that contain inconsonant figures (e.g., a modern automobile in a nineteenth-century scene); and (3) pictures that hide differences in detail. Artists have used the technique of embedding camouflaged figures in paintings—type (1)—for centuries, to add an element of surprise, bewilderment, or mischief to their visual texts. The nineteenth-century prints of the American lithographic company Currier & Ives featured hidden people, animals, and other objects. These held a broad appeal because they enticed viewers to detect the concealed images (Costello 1996: 51). The use of inconsonant figures—type (2)—also has a long tradition in painting. William Hogarth (1697–1764) and Norman Rockwell (1894–1978), for example, were well known for inserting inappropriate figures into their paintings, thus creating a sense of ambiguity and discordance that added considerably to the overall satirical effect of their art. Type (3) is a genre of puzzle known more generally as a *picture puzzle*. Solvers are given two pictures that seem identical but which, in fact, contain tiny differences:

> *There are five differences between Picture A and Picture B. Can you spot them?*

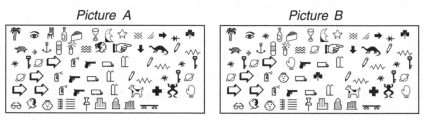

Figure 3.2

(See solution 3.1.)

As trivial as this type of puzzle appears to be, it nevertheless holds an important caveat, similar to the one built into a simple riddle. It warns us, in effect, to be wary of our visual perception, for we tend not to spot tiny differences between apparently similar things. Once they are spotted, however, the presence of such differences seems to create a momentary feeling of discord, upsetting our inborn need for visual harmony. By identifying those differences and marking them, we seem to be restoring harmony to the "look of things," albeit in a diminutive and inconsequential way.

Optical illusions further expose visual perception as basically unreliable. They do so largely in three ways, tricking the eye into (1) seeing images incorrectly; (2) interpreting images ambiguously; or (3) perceiving a figure or scene as representing something that is physically impossible. On the next page is an example of type (1) [see figure 3.3].

People reared in Western cultures typically see the line segment *AB* as longer than *CD*, even though the lines are equal in length. People reared in other cultures are not always fooled in this way. The illusion is called the Müller-Lyer Illusion, after the German physiologist Johannes Müller (1801–1858), who discovered it in 1840 (Rodgers 1998). The source

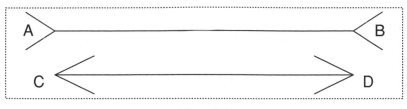

Figure 3.3

of the illusion is, clearly, the different orientations of the two arrow-heads, shapes that dupe the eye into perceiving one line as longer than the other. The use of arrowheads to distort the length of lines comes out of the tradition of perspective drawing, developed primarily by early Renaissance artists. It was systematized by Filippo Brunelleschi (1377–1446) and developed further by Albrecht Dürer (1471–1528). Perspective drawing fools the eye into seeing depth in a two-dimensional drawing. The Müller-Lyer Illusion is dramatic evidence that our eyes can be conditioned to see things not as they are, but as our representational systems want us to see them.

However, despite the fact that it was forged as a representational system aiming to deceive eyesight, perspective drawing has nevertheless been a very useful instrument in both art and geometry. In the case of the latter, it has provided tools for reasoning about space, as the preacher and literary critic Edwin A. Abbott (1838–1926) showed in his remarkable little novel of 1884, *Flatland: A Romance of Many Dimensions*. The characters of the novel are geometrical figures living in a two-dimensional universe called Flatland. Flatlanders see each other edge-on, and thus as dots or lines, even though from the vantage point of an observer in three-dimensional space looking down upon them they are lines, circles, squares, triangles, etc. To grasp the difference that the viewing perspective makes, think of Flatland as the surface of a table. If you crouch to look at a square piece of paper lying on a table, with your eyes level with the table's surface, you will see it as a line. The only way to see it as a square is to view it from above the table.

This type of "perspective thinking" has allowed geometers to entertain truly intriguing questions about space. For example: Is there a *formal* (i.e., "form-based") relation between our three-dimensional world—which can be called Sphereland—and the two-dimensional one? The answer is yes, because a Sphereland figure can be changed into a Flatland one, and vice versa, by making a specific kind of alteration to it. Consider a three-dimensional cube—a box made up of six sides, which are, geometrically speaking, six equal Flatland squares. This can be easily transformed into a Flatland figure, as shown below (with the art of perspective drawing, of course), by simply unfolding the six square sides onto a flat surface:

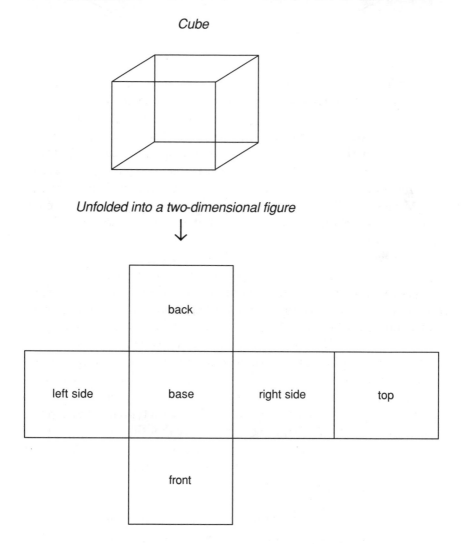

Cube

Unfolded into a two-dimensional figure
↓

Figure 3.4

The resulting Flatland figure can, of course, be just as easily transformed back into a three-dimensional cube by joining up the six squares. Now, the question that such folding and unfolding raises is truly mind-boggling: Can an analogous transformational method be envisioned that would produce a four-dimensional cube? Let's see (literally). We can start by drawing a Sphereland figure that corresponds to the unfolded Flatland figure above. It consists of eight equal cubes joined together, as shown below:

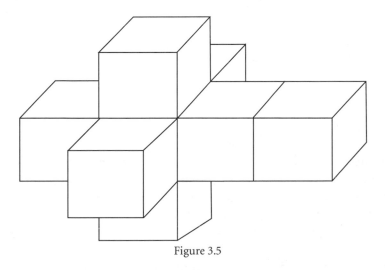

Figure 3.5

Now, the corresponding four-dimensional cube that presumably produced this Sphereland figure can be reconstructed by reversing the transformational process, i.e., by coming up with a figure that, when unfolded, would produce the eight-cube Sphereland figure above. The figure below will do so, as the more patient readers might wish to confirm for themselves:

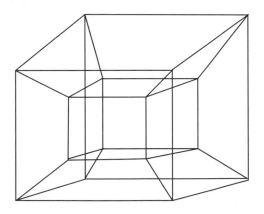

Figure 3.6

We now have before us—on two-dimensional paper, no less!—a drawing of what a four-dimensional figure might hypothetically look like. It is composed of 8 cubes, 16 vertices, 24 squares, and 32 edges. Geometers call it a *hypercube* or a *tesseract*. But is it truly a four-dimensional figure? Needless to say, there is no way to answer this question. Like the Flatlanders, we Spherelanders can only infer, through reasoning, what a four-dimensional world might look like. We will never be able to truly see one. Whether the hypercube really is a four-dimensional form or not in any meaningful physical sense is actually beside the point. The fact that it can be produced by pure imaginative-analogical reasoning is, in itself, the main lesson to be learned from Abbott's intriguing novel. While our viewing perspective may be limited by living in a three-dimensional space, our imagination is not. Incidentally, Pappas (1999: 23) credits the nineteenth-century Swiss mathematician Ludwig Schläfli with the first drawing of the hypercube—a drawing that aroused the imagination of modern-day artists, such as the great Spanish surrealist painter Salvador Dalí (1904–1989), whose striking *Crucifixion* (1954) shows Christ on an unfolded hypercube cross.

Type (2) optical illusions, known in psychology simply as *ambiguous figures,* are drawings that our eyes see as different figures at different times (Luckiesh 1965). For example, the following picture—created by a psychologist named Joseph Jastrow around 1900—may be seen by our eyes as either a duck or a rabbit (Block and Yuker 1992: 16):

Figure 3.7

The illusion on the facing page—devised by the Dutch psychologist Edgar Rubin around 1910—produces the same type of "double-vision" effect. At one time, it appears to be a vase, and at another, the faces of two people. Rubin suggested that the ambiguity resulted from the brain's inability to make a distinction between a shape in the foreground and a shapeless background (Rodgers 1998: 40):

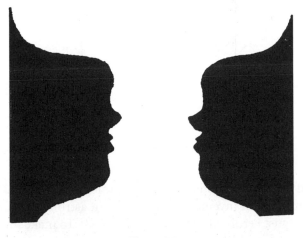

Figure 3.8

One of the most famous of all ambiguous pictures was created by none other than Salvador Dalí, who duped the eye into seeing the faces of two nuns and the bust of Voltaire in his painting *Slave Market with the Apparition of the Invisible Bust of Voltaire*. Type (2) optical illusions show, as Shepard (1990: 168) aptly puts it, that our "visual experience evidently is the product of highly sophisticated and deeply entrenched inferential principles that operate at a level of our visual system that is quite inaccessible to conscious introspection or voluntary control." The ways in which our eyes interpret three-dimensional space impel us to interpret flat figures on surfaces in the same ways. Such interpretation is the source of type (2) illusions.

Pictures or objects that appear to contradict physical laws—type (3) illusions—have a jarring effect upon us. One of the first to be studied scientifically was the Zöllner Illusion, discovered by astrophysicist Johann Zöllner (1834–1882), who stumbled upon a piece of fabric with a design that made parallel lines appear decidedly unparallel to the eye. The four lines on the next page are parallel, but they do not seem to be so. They are an example of the Zöllner Illusion [see figure 3.9].

The source of the illusion is, clearly, the small slanted lines on the parallel lines. Once again, we interpret the lines as unparallel because we are conditioned to see drawings on two-dimensional surfaces in three-dimensional perspective.

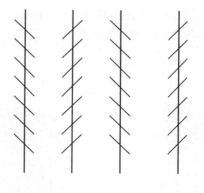

Figure 3.9

One of the most famous of all three-dimensional type (3) illusions is the Möbius Strip, invented by the German astronomer and mathematician August Ferdinand Möbius (1790–1868). To make such a figure, draw a dotted pencil line down the middle of a flat rectangular strip of paper:

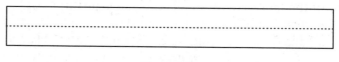

Figure 3.10

Then give the strip a half-twist and join the ends:

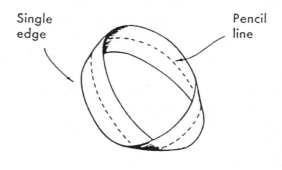

Figure 3.11

Now, how many sides does this strip have? Running a pencil along the dotted line brings us right back to where we started. Thus, it would appear that the strip has only one side, although the original unjoined strip had two! Even more perplexing is the fact that if one cuts the

Möbius Strip in two along the pencil line, it does not come apart. As if by magic, two strips linked together are produced—as readers can verify for themselves. The German mathematician Felix Klein (1849–1925) became so captivated by the Möbius Strip that he invented a "bottle" version of it, known appropriately as the Klein Bottle. The bottle is one-sided. It is a closed shape with no ends, yet it has no inside. Indeed, if water were poured into it, the water would come out of the same hole into which it was poured. If cut in two lengthwise, the bottle forms two Möbius strips:

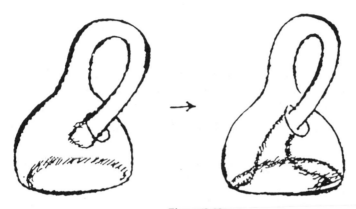

Figure 3.12

Such oddities defy explanation. The Dutch artist Maurits Escher (1898–1972) became widely known for incorporating them into his paintings. Beginning about 1936, Escher started drawing ambiguous patterns in which he interlocked repeated figures of stylized animals, birds, or fish, leaving no spaces between the figures (Schattschneider 1990). A little later, he began toying with the viewer's perceptions, creating such illusions as staircases that appeared to lead both upward and downward in the same direction, and alligators that seemed to come to life, walking off the edge of the paper. The essence of Escher's technique can be seen in the drawing on the next page, which is imitative of the kind of staircases Escher often drew [see figure 3.13].

In this two-dimensional picture there appears to be no highest or lowest step. For example, if one starts climbing at D, moving counter-clockwise, one ends up back at D, having apparently moved upward with each step and yet ending no higher than before. Similarly, if one moves clockwise, descending from D, one ends up, again, at D. Such a figure appears, therefore, to contradict all the principles of physics. Per-

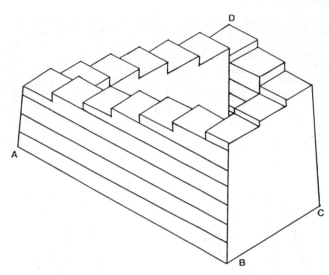

Figure 3.13

haps the only way to grasp this illusion in rational terms is to "slice the staircase," as Falletta (1983: 32) proposes. Doing so shows that the levels of the staircase do not lie in a horizontal plane, but rather move upward spirally. The steps of the staircase, on the other hand, remain in the horizontal plane. But Falletta's proposal still does not explain why we end up perceiving it as we do when it is unsliced!

One of the most prolific producers of this kind of art is the Swedish artist and art historian Oscar Reutersvärd (b. 1915). His drawings have captured the attention of mathematicians and psychologists alike, who use them as templates for studying visual perception and representations of three-dimensional space. Here is one of his most famous works, which the psychologist Hoffman (1998: 4) calls the "devil's triangle" because it plays fiendishly on our fallible visual interpretation, creating a jarring sense of distortion and surreal unease:

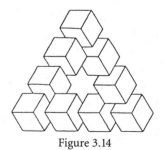

Figure 3.14

In 1958, the English biologist L. S. Penrose and his son Roger drew their own version of the devil's triangle:

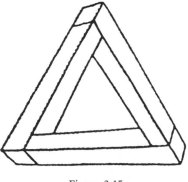

Figure 3.15

Such oddities have been important in the development of *topology*, which, in turn, has had many implications for the study of Nature (Fauvel, Flood, and Wilson 1993). Topology is the branch of geometry that investigates the relations of figures and shapes in spaces of various kinds. The first systematic treatment of this science appeared in 1847 in a work by a German mathematician named Listing, although the word "topology" was coined much later (in 1930) by another mathematician, Solomon Lefschetz. Topology concerns itself with determining the *insideness* or *outsideness* of shapes. A circle, for instance, divides a flat plane into two regions, an inside and an outside. A point outside the circle cannot be connected to a point inside it by a continuous path in the plane without crossing the circle's circumference. If the plane is deformed, it may no longer be flat or smooth, and the circle may become a crinkly curve, but it will continue to divide the surface into an inside and an outside. A knot, too, may be thought of as a simple closed curve that can be twisted, stretched, or otherwise deformed, but not torn (Neuwirth 1979). Two knots are equivalent if one can be deformed into the other; otherwise, they are distinct.

Geometrical Mind-Bogglers

Optical illusions underline both how unreliable visual perception is and how restricted Euclidean geometry turns out to be for investigating our physical world, with all its topological peculiarities. Euclidean geometry has, however, provided much fodder for a host of clever puzzlists throughout history. More than any other principle within that system of geometry, the Pythagorean Theorem has been the inspiration for innumerable ingenious puzzles. The following one, for instance, was devised by the Indian astronomer and mathematician Bhaskara (1114–c. 1185) in his famous work, the *Lilavati* (Wells 1992: 118–119). Although it is a relatively simple puzzle, it is worth discussing in detail here because it illustrates rather nicely how puzzlists can produce a mind-boggling effect with a basic geometrical principle:

> *A snake's hole is at the foot of a pillar which is 15 cubits high, and a peacock is perched on its summit. Seeing the snake, at a distance of thrice the pillar's height, gliding toward his hole, the peacock stoops obliquely upon him. Say quickly, at how many cubits from the snake's hole do they meet, both proceeding an equal distance?*

What makes this puzzle mind-boggling for many solvers at first reading is the fact that neither the peacock's oblique attack on the snake nor the location of their meeting point is immediately visualizable. The key to solving this, and all similar puzzles, is to draw a diagram that will visually represent the information presented by the puzzle. First, we can draw the pillar as the vertical line PQ, labeling it 15 cubits high—the top point P representing the perch on which the peacock sits. Then we can draw a line from the pillar's base (Q), and at a right angle to it, to a point S, representing the snake's initial position. The line QS can be labeled 45 cubits long, for this is "a distance of thrice the pillar's height" ($15 \times 3 = 45$):

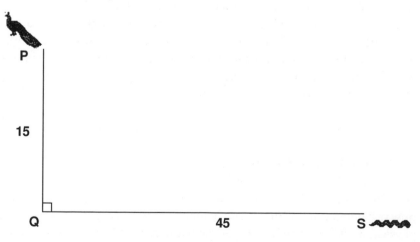

Figure 3.16

We are told that the snake is moving along the ground toward the pillar and that the peacock stoops on the snake obliquely. All we have to do is mark a point on the line QS, calling it R, where the peacock meets the snake. Both animals, Bhaskara tells us, travel an equal distance to point R, although their paths are at an unknown angle to each other. Therefore, x can stand for both the distance from P (the peacock's perch) to R and the distance from S (the snake's starting point) to R, and we can label the two line segments accordingly. We can then label the distance from the point where the two meet (R) to the pillar's base (Q) $(45 - x)$, the total distance between the pillar and the snake minus the distance already covered by the snake when it reaches R:

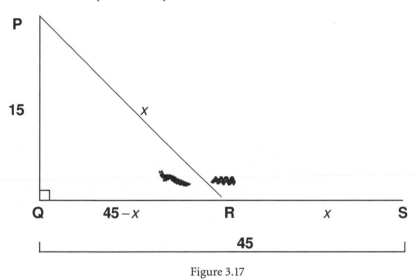

Figure 3.17

The diagram now makes it obvious that the Pythagorean Theorem can be applied to the right triangle PQR in order to determine the length x. This produces the equation $x^2 = 15^2 + (45 - x)^2$. Solving it, we get $x = 25$. Therefore, the distance QR, or $(45 - x)$, is $45 - 25 = 20$. The peacock thus meets the snake at a distance of 20 cubits from the base of the pillar.

The salient characteristic of this puzzle is that it seems initially to be unsolvable. Bhaskara accomplished this "mind-boggling" effect primarily by playing on the unknown angle of the paths traveled by the two creatures. Other puzzlists have induced the same kind of effect by presenting, instead, diagrams that do not seem to contain enough information to permit a solution. The following is an example of this second type of mind-boggler. To the best of my knowledge, it was first devised by Martin Gardner (e.g., 1994a: 14):

> *Given the length of OC as 8 and CD as 2, and the fact that O is the center of the circle, figure out the length of the diagonal AC of the rectangle AOCB.*

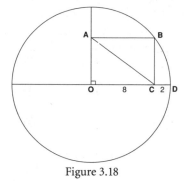

Figure 3.18

Many people look at this diagram, try out a series of complicated calculations using the Pythagorean Theorem, and end up obtaining only frustration for their efforts. As it stands, Gardner's diagram simply appears to have insufficient information in it. But therein lies the trick—the diagram is, in fact, incomplete. The solution becomes instantly visualizable with the addition of the other diagonal, OB:

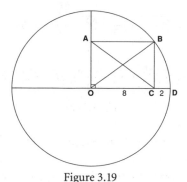

Figure 3.19

Now it can be seen that OB and OD are radii of the circle, and we know that radii of the same circle are all the same length. OD is 8 + 2 = 10 units long. Therefore OB is also 10 units long. Since we also know that the diagonals of a rectangle are equal, OB = AC = 10. As was the case with

Bhaskara's puzzle, the striking feature of Gardner's mind-boggler is that it dupes the solver into thinking that the puzzle is much more difficult or confusing than it actually is.

The mind-boggling effect is also produced by the Nine-Dot Four-Line Puzzle, which has become a classic in this genre:

> *Without the pencil leaving the paper, can four straight lines be drawn through the following nine dots?*

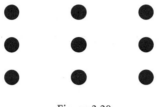

Figure 3.20

People typically approach this puzzle by attempting to join up the dots as if they were located on the perimeter of a 3 × 3 rectangle (Sternberg and Davidson 1982: 38–39):

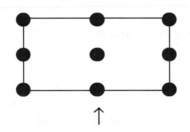

Nine dots seen as being on the perimeter of a 3 x 3 rectangle

Figure 3.21

As readers can verify for themselves, no matter how many times one tries to join the nine dots with four lines, one or two dots are always left over. However, if the dots are not viewed as being on the perimeter of a square, a simple solution presents itself to the eye. First, we start by putting the pencil on, say, the bottom left dot, tracing a straight line upward and stopping when it is in line diagonally with the two dots below it, as shown. We could start with any of the four corner dots and produce the solution (as readers can discover for themselves) [see figure 3.22].

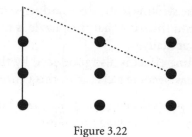

Figure 3.22

Next, we join the two dots (without raising the pencil) by tracing a straight line diagonally downward through them, stopping when our second trace is in horizontal alignment with the next three dots, as shown.

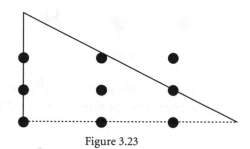

Figure 3.23

The third line can now be drawn horizontally, from right to left, joining the three dots on the bottom row.

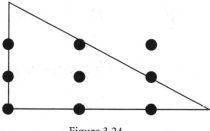

Figure 3.24

Finally, with a fourth line, the two remaining dots can be joined [see figure 3.25].

As this puzzle demonstrates, the eye can be easily duped into assuming something that is not necessary—namely, that the dots are on the perimeter of an imaginary rectangle. As we saw in the discussion of optical illusions, this puzzle once again warns us that perception is

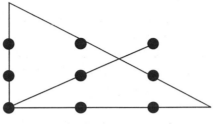

Figure 3.25

shaped by experience. In our culture, we are conditioned to see the rectangular form not only in actual rectangular figures, made up of four lines meeting at right angles, but also in detached dots (or other marks) that appear to trace the outline of a rectangular figure.

Geometrical mind-bogglers require an abundant use of insight thinking, rather than a straightforward application of accurate reckoning. Once the necessary insight is attained, they lose their fascination. This brings to light, in microcosm, the reason that mathematical discoveries are organized into categories. Indeed, the whole process of organizing mathematical knowledge might be said to consist in the effort to minimize the need to repeat insightful thinking. Once an insight is attained, it becomes useful to routinize it, so that a host of related problems can be solved as a matter of course, with little time-consuming mental effort. Such routinization is a memory-preserving and time-saving strategy. It is the rationale behind all organized knowledge systems. Such systems produce *algorithms*—routinized procedures—for solving problems that would otherwise require insight thinking to be used over and over again. Once such thinking has done its job, so to speak, the rational part of the mind steps in to give its products form and stability through organization.

No wonder, then, that mathematicians are so averse to the existence of gaps within their corpus of algorithmic knowledge, never giving up on a problem until it is solved or proved to be truly unsolvable. A case in point is the background story of a famous geometrical mind-boggler known as the Four-Color Problem. Mapmakers had believed from antiquity that four colors were sufficient to color any map so that no contiguous regions would share a color. This insight caught the attention of mathematicians in the nineteenth century, after a young mathematician at University College, London, named Francis Guthrie (1831–1899) formally proposed in 1852 that four colors would always be enough, although there is some evidence that Möbius had proposed it in a lecture to his students as early as 1840 (Falletta 1983: 159). Guthrie apparently wrote about the problem to his younger brother, Frederick. The story then goes that Frederick described the problem to his own professor, the prominent British mathematician and puzzlist Augustus De Morgan (chapter 1). De Morgan quickly realized that the Four-Color Problem had many important ramifications for mathematical method. Word of the problem spread quickly.

The following map, devised by Lewis Carroll, is a simple yet elegant portrayal of the problem. It shows four countries, each of which shares boundaries with the other three. The map cannot be colored with fewer than four colors so that no two regions share a boundary of the same color. The map below uses blue, red, green, and yellow for the sake of concreteness. But a combination of any four colors would do just as well.

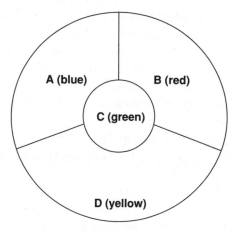

Figure 3.26

As can be seen, region A is touched by red, green, and yellow; region B by blue, green, and yellow; region C by blue, red, and yellow; and region D by blue, red, and green. These four colors are sufficient to ensure that no two regions sharing the same boundary are touched by the same combination of colors.

Carroll demonstrated that there are maps that cannot be colored correctly with three colors. But is it true, as mapmakers believed for centuries, that four colors are always enough, for any map? For over a century after Guthrie articulated the Four-Color Problem, mathematicians attempted to demonstrate its validity with the usual deductive QED mind-set. But their efforts proved to be consistently fruitless. In 1976, instead of using the QED method, mathematicians Wolfgang Haken and Kenneth Appel finally solved it at the University of Illinois, with the aid of a computer (Haken 1977; Haken and Appel 1977). The program they wrote was thousands of lines long and took 1200 hours to run. Since then, mathematicians have been checking the program, finding only minor and fixable problems in it (Devlin 1998b: 148–176). Haken and Appel essentially showed that no exception to the four-color conjecture can be found. However, to this day many mathematicians remain uncomfortable with their demonstration, for the simple reason that it is clearly not based on the method of deductive demonstration. Soon after the publication of the Haken-Appel proof, Thomas Tymoczko (1979) encapsulated the feelings of many when he observed that, if accepted, Haken and Appel's work puts mathematics in a position to radically alter its traditional view of proof. The same observation applies, incidentally, to the recent proof of Fermat's famous theorem, as we shall see in the final chapter. The American philosopher Thomas Kuhn (1922–1996) claimed that scientists work within a paradigm (a set of accepted beliefs), which eventually weakens until new theories and scientific methods replace it. By accepting a new way of proving the four-color theorem (albeit grudgingly), mathematics does indeed seem to be undergoing a paradigm shift, as Tymoczko suggests.

Puzzles in Geometrical Dissection and Arrangement

The dissection and rearrangement of geometrical figures has intrigued mathematicians since antiquity. Euclid devoted a large portion of his *Elements* to questions of how to dissect angles and lines, and how to construct certain figures. The Greek geometers were also responsible for three of the most famous of all puzzles in the domain of geometry. They wanted to know if it was possible, using compass and straight-edge alone, to (1) trisect an angle into three congruent angles; (2) construct a cube with twice the volume of a given cube; and (3) construct a square with the area of a given circle. These three problems stimulated mathematical thought and work for over two thousand years, until mathematicians showed in the nineteenth century that they were not solvable with compass and straight-edge alone.

The fascination with dissection and rearrangement has also been the source of countless solvable puzzles across history (Lindgren and Frederickson 1972). Like mind-bogglers, these depend largely on the resources of the imagination for their solution. Here is an example of one such puzzle in this genre, invented by a certain Angelo John Lewis (who really was Professor Hoffman) and found in his 1893 puzzle compilation, *Puzzles Old and New:*

> *How can the following rectangle, with its two tabs, be cut into two pieces to make a complete rectangle?*

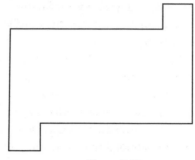

Figure 3.27

This puzzle invariably dumbfounds solvers initially. What kind of cut would produce two pieces which will form a rectangle without the tabs? The appropriate *Aha!* insight is cutting the figure in a zigzag fashion, with each cut equal in length to a side of the tab (see solution 3.2).

The above puzzle is essentially a *disappearance puzzle,* since the end result of the zigzag cut is the disappearance of the two tabs. Other puzzlists have used the same kind of "cut and paste" principle to create highly mystifying puzzles. For example, in his *Rational Recreations* of 1774, a certain William Hooper posed the *vanishing puzzle,* whose central idea was subsequently adopted and made famous by no other than Sam Loyd in what was called the Disappearing Warrior puzzle. Loyd created his version by fastening a smaller paper circle to a larger one with a pin so that it could spin around. Then with appropriate artwork on both circles, he made the figure look like the earth with thirteen warriors on it (Gardner 1982: 64–65) [see figure 3.28].

Figure 3.28

When the smaller circle is turned slightly clockwise, as shown below, the thirteen warriors turn mysteriously into twelve. Where did the thirteenth warrior go?

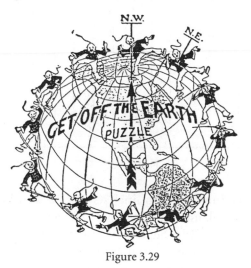

Figure 3.29

Loyd patented his puzzle in 1896. It sold more than ten million copies. To grasp its principle, which he adapted from Hooper's original creation, consider the following diagram, consisting of ten equidistant parallel lines, with a dotted diagonal (from Gardner 1982: 68):

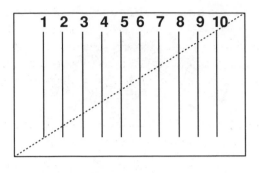

Figure 3.30

You can draw this figure on a piece of paper. Cutting the paper along the dotted line produces an upper and a lower piece. Slide the lower one down and to the left, until the lines are realigned as shown. This produces nine lines, since line 10 has become an edge:

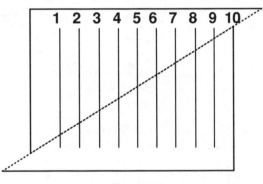

Figure 3.31

What happened to the tenth line? As can be seen in the diagram above, the ten original lines are sliced by the diagonal as follows: lines 1 and 10 are untouched, while the remaining eight lines (2 to 9) are sliced into two pieces each. The total number of lines after the cut is therefore $2 + 16 = 18$. The slide matches them up in pairs, realigning the eighteen lines into nine. Of course, each of the lines is longer by $\frac{1}{9}$ than it was before the slide. The apparent disappearance of a line is due to the realignment caused by the slide. Indeed, if you slide the lower part back up again, the tenth line will reappear. This is, essentially, the principle used by Hooper and adapted by Loyd to create his mystifying Warrior puzzle. Without going into details here, suffice it to say that when one of Loyd's two circles is turned, the body parts of the warriors on the circles are realigned, making it seem that one has disappeared.

One of the cleverest arrangement puzzles ever devised comes from the pen of the medieval Persian mathematician Abu Wafa (A.D. 940–998), a well-known commentator on the works of Diophantus and Al-Khwarizmi:

> Draw three identical triangles, and one smaller triangle similar to them in shape, so that all four can be made into one large triangle.

The puzzle as stated fools many solvers into thinking that the larger triangle should enclose all four triangles. But its statement does not ask us to do that. Abu Wafa simply put three larger triangles, 1, 2, and 3 around a smaller one, 4, and then joined three vertices (as shown), thus producing one large triangle (shown by the dotted lines):

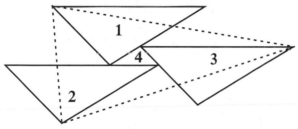

Figure 3.32

Just as ingenious as Abu Wafa's brainteaser is the following contemporary puzzle posed by Martin Gardner (1994a: 16–17):

> Given an obtuse triangle (a triangle with an angle greater than 90°), is it possible to cut the triangle into smaller triangles, all of them acute (triangles with all of their angles less than 90°)?

Gardner points out that attempts to solve this problem using "linear logic" lead nowhere, as with the nine-dot puzzle above. A little imagination, rather than simple linear reasoning, however, will yield the *Aha!* insight (see solution 3.3).

Not all dissection puzzles rely so extensively on insight thinking for their solution. One of the classic puzzles in this genre, as a matter of fact, primarily requires accurate reckoning:

> With one cut, a perfectly circular pie can be cut into two pieces. With a second cut that crosses the first one, four pieces will be produced. With a third cut, seven pieces can be produced:

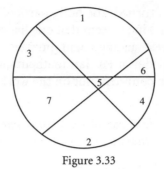

Figure 3.33

What is the largest number of pieces that can be produced with six straight cuts?

We start by considering simpler cases, in a logical manner, in order to see if a general pattern can be extracted from them. First, we slice the pie with one cut, which of course generates two pieces.

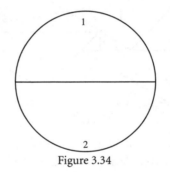

Figure 3.34

With a second cut, we can get at most four pieces.

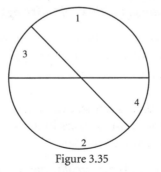

Figure 3.35

A third cut generates at most seven pieces.

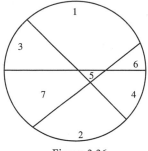

Figure 3.36

We can tabulate the results so far as follows:

Number of Cuts	Maximum Number of Pieces
1	2
2	4
3	7

Is there a pattern? It seems that each new cut adds a number of pieces equal to the number of the cut. For instance, if the number of cuts is 1, then the number of new pieces is going to be 1. So, adding the number of cuts (which is the number of new pieces) to the number of pieces already existing, we get $1 + 1 = 2$ pieces after cut 1. If the number of cuts is 2, then the number of new pieces is going to be 2. Adding this number to the previous number of pieces, we get $2 + 2 = 4$ pieces. If the number of cuts is 3, then the number of new pieces is going to be 3. Adding three new pieces to the previous number of 4, we get $3 + 4 = 7$ pieces. This *algorithm,* or method of computing the number of pieces from the number of cuts, predicts that a fourth cut should add 4 new pieces to the previous 7, thus yielding $4 + 7 = 11$ pieces:

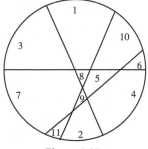

Figure 3.37

As predicted, it does. Filling in our chart up to 6 cuts using the above algorithm shows that 22 pieces of pie will be produced with 6 cuts:

Number of Cuts	Number of Pieces
1	2
2	4
3	7
4	11
5	16
6	22

Note that we cannot say for sure that this is the true solution until we have actually tried it out on a circle figure. This is because the method used to reach the solution was *inductive:* i.e., it involved using individual facts to infer a general rule or pattern. On the other hand, the method used by Euclid to prove the equality of the vertically opposite angles produced by two intersecting lines was *deductive:* i.e., it involved reasoning from known premises and facts to a logically valid conclusion. As we shall see in the next chapter, deduction is the heart and soul of mathematical and logical systems. Deduction demonstrates that something is the way it is beyond any shadow of a doubt; induction from a set of facts, on the other hand, can only suggest that something is the way it is, not demonstrate it. No matter how many cuts we make to the circle, we can never be sure that the above algorithm is true in every case. Hence, conclusions drawn from inductive reasoning are said to be probably true, but not always or necessarily true. Glenn and Johnson (1961: 11–12) provide a useful illustration of why inductive reasoning is not always reliable. Consider the following arithmetical computations—multiplications are on the left and additions on the right.

Multiplication	= Addition ?
$2 \times 2 = 4$	$2 + 2 = 4$
$\frac{3}{2} \times 3 = 4\frac{1}{2}$	$\frac{3}{2} + 3 = 4\frac{1}{2}$
$\frac{4}{3} \times 4 = 5\frac{1}{3}$	$\frac{4}{3} + 4 = 5\frac{1}{3}$
$\frac{5}{4} \times 5 = 6\frac{1}{4}$	$\frac{5}{4} + 5 = 6\frac{1}{4}$

From these examples, we might be led to induce that multiplying numbers always produces the same result as adding them. But of course, nothing could be further from the truth.

The great Archimedes was particularly adept at experimenting with dissection and arrangement patterns. His Loculus is a dissection puzzle in which a square is cut into fourteen pieces that are to be reassembled to form silhouettes of people, animals, or objects. Archimedes' original puzzle is lost; the version that we have today comes down to us from an Arabic manuscript titled *The Book of Archimedes on the Division of the Figure Stomaschion*:

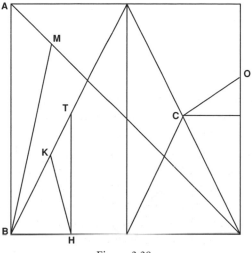

Figure 3.38

The points M, T, and C are midpoints of line segments; HK would pass, if extended, through A, and OC, if extended, through B. The fourteen pieces this figure is divided into can be used to make various shapes (animals, buildings, etc.). Readers can verify this for themselves by making a cardboard version of the Loculus.

The Loculus has never really caught on with puzzle enthusiasts. On the other hand, a Chinese version of Archimedes' puzzle, known as the *tangram*, has become highly popular in all parts of the world. Its origin is unknown. Some believe that the word "tangram" is of American origin, derived from a Chinese slang word for "prostitute" (*tan*), claiming that American sailors who frequented Chinese prostitutes were entertained not only sexually by the women, but also intellectually with such puzzles. Others believe that the tangram originated in ancient China (Vorderman 1996: 130). As Slocum and Botermans (1992: 8) suggest, however, the most likely source of the tangram puzzle is the late-eighteenth-century Chinese puzzle Ch'i Ch'io, invented during the rule (1796–1820) of emperor Chia Ch'ing. But the mathematician Takagi disagrees (1999). He believes the tangram is of Japanese origin, because he found a remarkable little book containing seven-piece tangram puzzles, titled *The Ingenious Pieces of Sei Shonagon,* that was published in 1742 in Japan.

Whatever the real origin of the tangram, the puzzle made its way to Europe and America in the early nineteenth century, where in 1817–1818 it became the first true international puzzle craze. It is said that even Napoleon was an avid tangram player while in exile on St. Helena. Lewis Carroll, Sam Loyd, and Henry Dudeney, among other puzzlists, created many ingenious tangram puzzles. Sam Loyd devoted an entire book to this puzzle genre, *The Eighth Book of Tan,* when he was sixty-one, to show his huge audience of readers how truly intriguing it was (Loyd 1968).

There are seven tangram pieces—five triangles, one square, and one parallelogram. These are constructed by cutting pieces from a square template:

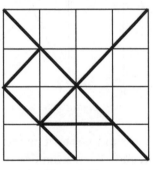

Figure 3.39

The seven pieces may be cut out as many times as needed from multiple templates, producing as many triangles, squares, and parallelograms as the puzzle calls for. The challenge lies in rearranging them to make new patterns or figures (Read 1965). For example, two parallelograms, one square, and a trapezoid figure (made with triangles and a square) can be rearranged to suggest the figure of a rabbit:

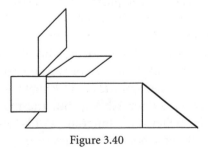

Figure 3.40

Why are such rearrangement puzzles so pleasing? In my view, we are fascinated by them because we are fascinated by pattern and form. For Pythagoras, the harmonious arrangement of things in the universe was captured by the symmetry of geometrical figures. This is why, he claimed, the human eye derives such great pleasure from seeing geometric patterns. Some specific figures are especially pleasing. One is the *golden ratio* or *section*. This can be represented geometrically as follows. We begin by drawing a rectangle with sides in the proportion 1:2. The diagonal is, according to the Pythagorean Theorem, √5:

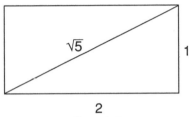

Figure 3.41

The golden ratio, often written as ϕ, is $(1 + \sqrt{5})/2$, a number mentioned at the beginning of Book VI of Euclid's *Elements*. It has the numerical value 0.6180339. . . . Since antiquity many philosophers, artists, and mathematicians have been intrigued by the golden ratio, which Renaissance writers called the *divine proportion* (Huntley 1970). It is widely accepted that a figure or form incorporating this ratio exhibits a special beauty. Incidentally, the Egyptians also used the golden ratio, which the Rhind Papyrus refers to as a "sacred ratio," in building the Great Pyramid at Giza. Artists and architects throughout history (Pedoe 1976) have employed this ratio. The rectangular face of the front of the Parthenon in Athens has sides whose ratio is golden. The ratio of the height of the United Nations Building in New York (built in 1952) to the length of its base is also golden (Kappraff 1991). Remarkably, the ratio of two successive terms in the Fibonacci sequence (chapter 1)—1, 1, 2, 3, 5, 8, 13, 21, 34 . . . —converges on the golden ratio: e.g., $\frac{5}{8}$ = .625, $\frac{8}{13}$ = .615, $\frac{13}{21}$ = .619, etc. It has also been found in Nature, such as in the curve of a nautilus shell. One cannot help but be struck by such serendipities. As Pappas (1999: 5) appropriately puts it, this ratio mysteriously "sets off a truly amazing chain of mathematical interconnections."

Pythagoras's belief that geometry expresses an inherent order in the universe has, curiously, received some modern-day corroboration in the intriguing work of the mathematician Benoit Mandelbrot (b. 1924). Mandelbrot found that random fluctuations in Nature and in human affairs form geometrical patterns, which he called *fractals,* when they are reduced to smaller elements. A fractal is defined as any form that is altered by application of a *transformational rule* to it ad infinitum. Coined from the Latin word *fractus,* Mandelbrot's term suggests fragmented, broken, and discontinuous phenomena. But, as it turns out, fractals disclose a strange type of hidden pattern in shapes that would otherwise appear random to the naked eye. Here's an example of a fractal, known as the Sierpinski triangle, named after the Polish mathematician Vaclav Sierpinski (1882–1969) (Barnsley 1988):

Step 1:   *Draw an equilateral triangle:*

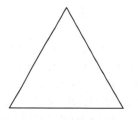

Step 2:   *Divide it into three congruent triangles (note that this generates a fourth triangle, with reverse orientation to the other three):*

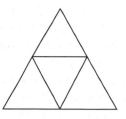

Step 3:   *Repeat this process on the figure produced by step 2 ad infinitum:*

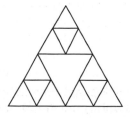

Figure 3.42

The reader may legitimately ask, what is the point of such a play on geometrical form? Remarkably, snowflakes have a fractal form. Like the Sierpinskian triangle, a snowflake seems to be generated by Nature with a transformational rule—a connection discovered by Swedish mathematician Helge von Koch (1870–1924) in 1904:

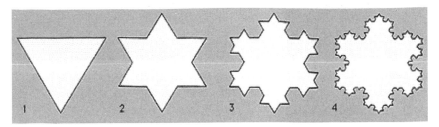

Figure 3.43

Fractal shapes were known to the human imagination long before fractal geometry provided a theory for them. They turn up in the ancient world, in Islamic art, and in Celtic artifacts. An early prototype of the Sierpinski figure crops up in a thirteenth-century pulpit in the Ravello cathedral in Italy, designed by Nicola di Bartolomeo of Foggia (1230–1272). In Mahayana Buddhism, the fractal nature of reality is captured in the Avatamsaka Sutra by the god Indra's net, a vast network of precious gems hanging over Indra's palace, so arranged that all the gems are reflected in each one. In recent times, both Dalí and Escher have exploited the fractal technique of creating a new shape out of repeated copies of another, as has the abstract expressionist painter Jackson Pollock (1912–1956).

Amazingly, it turns out that fractals are perfect models for understanding the shapes of such things in Nature as fern leaves, snowflakes, lava flows, coastlines, and mountain terrains (Barnsley 1988; Devlin 1997), confirming not only Pythagoras's belief that geometry is a model of Nature, but also the great scientist Galileo Galilei's (1564–1642) observation that the book of Nature can be read only by those who know the language in which it was written—mathematics. Fractal geometry is, clearly, a modern-day dialect of that language.

Mazes

Labyrinths are essentially fractal designs, seemingly chaotic, yet concealing a hidden pattern. The puzzle genre derived from the concept of the labyrinth is known as the *maze*. It is a network of intricately winding paths that appear to have no order whatsoever. The challenge lies in following—and, perhaps, in finding—the one path that will lead either to the center or to a way out.

As mentioned in the opening chapter, labyrinths are intertwined with mythic traditions. The labyrinthine prison on Crete, for instance, was constructed by the Athenian craftsman Daedalus for King Minos of Crete, to detain the half-human, half-bull Minotaur. According to legend, around 1600 B.C. Androgeus, the king's son, was killed, apparently by Athenians. Meanwhile, his wife Pasiphae had fallen in love with a bull and given birth to the Minotaur. Shamed by this event and aching to exact his revenge on the Athenians, Minos exacted a tribute of seven young men and women every year, whom he sent into the prison. At its center he put the abominable Minotaur, a beast eager to destroy anyone who arrived. Theseus, son of King Aegeus of Athens, offered to go as one of those to be sacrificed. Minos's clever daughter, Ariadne, who was in love with Theseus, gave her beloved a sword with which to kill the Minotaur and a thread to mark his path through the maze. Theseus slew the Minotaur and emerged to reunite with Ariadne, finding his way back by simply following the thread. Aegeus had instructed Theseus to raise a white sail on his ship if he had accomplished his mission. But Theseus forgot to do so and, as legend has it, when his father saw the ship returning with black sails, he threw himself into the sea, which was thereafter called the Aegean.

The Cretan labyrinth has appealed to rulers, philosophers, mathematicians, artists, and writers alike. The later Roman emperors had copies of the labyrinth embroidered on their robes. Cretan labyrinths can be seen etched on the walls of many early Christian churches. The surrealist Argentine writer Julio Cortázar (1914–1984) was so taken by the story of the labyrinth that he portrayed the outside world in his novels as a phantasmal Cretan maze from which every human being must escape. Cortázar's great contemporary and compatriot, Jorge Luis Borges (1899–1986), was also apparently spellbound by the labyrinth, writing a series of truly intriguing stories collected under the title *Labyrinths*. As the work of these two great writers suggests, the labyrinth concept constitutes an unconscious metaphor for the mystery of life, which is felt to inhere in a series of seemingly random paths that conceal a hidden pattern leading to some secret center which may hold the solution to the mystery.

The oldest labyrinthine design found is carved into the stone wall of a five-thousand-year-old grave in Sicily. Similar carvings have been discovered throughout the world. The greatest number, over three hundred, have been found in Sweden and in Gotland in the Baltics (Gullberg 1997: viii). Labyrinthine patterns have also been found on stone carvings in Ireland that go back to around 2000 B.C., in the Alps, at Pompeii, in Scandinavia, Wales, England, Africa, and in Hopi Indian rock carvings in Arizona (Fisher and Gerster 1990: 32–44). Many ancient buildings and cities were designed as labyrinthine structures. The Egyptian pyramids and the Christian catacombs—the networks of subterranean chambers and galleries used for burial by peoples of the ancient Mediterranean world—were designed as labyrinths, presumably to test the ability of the deceased to figure out the right path to the afterworld (Hooke 1935; Lockridge 1941). The Egyptians also built various buildings as labyrinths. The largest one was the Great Labyrinth, a huge building constructed around 2000 B.C. in northern Egypt, with three thousand rooms. The ancient city of Troy was designed with labyrinthine paths, providing protection against invaders by confusing them (Gullberg 1997: viii).

The labyrinth design reverberates with mystical connotations. In Java, Sumatra, and India, this design has been used from time immemorial as a tool for achieving inner peace. The Navajo people in the United States see the labyrinth as a representation of how the world was created. The floors of some medieval churches had labyrinthine designs in them to symbolize the tortuous journey of individuals toward salvation (Doob 1990: 18). One of the largest can be found at Chartres, where the faithful would make pilgrimages on their knees through it. Readers can trace their own way through the Chartres maze [see figure 3.44].

In Renaissance Europe labyrinths were added to murals and pavements, and many gardens were designed as mazes walled by clipped hedges. Two of the best known were built in the seventeenth century: the maze garden at Hampton Court, in London, and the exquisite labyrinth in the Palace of Versailles, probably built by Louis XIV (1638–1715), which is adorned by thirty-nine fountains and various statues depicting characters from Aesop's fables (Mathews 1970). Such labyrinths share a basic design [see figure 3.45].

The construction of such gardens continues uninterrupted to this day. In the United Kingdom, a company called Minotaur Designs (Costello 1996: 70) constructs landscape mazes for anyone interested in landscaping with a special motif. The most famous maze built by the company is the Beatle Maze in Liverpool, completed in 1984, in honor of the Beatles. It is dominated at its center by a yellow submarine, after their famous song. The maze itself is in the shape of a pair of ears, with mu-

Figure 3.44

sical notes surrounding it. The path consists of four parallel rows of bricks, corresponding to each of the four members of the original rock group. When it passes the statue of John Lennon, it diminishes to three rows, in memory of his tragic death.

Not surprisingly, the labyrinth concept has also been the source of many puzzles throughout the ages. Here is an interesting modern-day

Figure 3.45

puzzle version of the Cretan labyrinth, adapted from David Wells (1992: 118):

> *What is the shortest route that can be taken for avoiding the Minotaur, if, in order to avoid an ambush from the beast from one of the many passages in the labyrinth, you must turn left or right (in a zigzag fashion)? Like Theseus, you must enter by the south entrance to the palace, wanting to get to Ariadne, who is waiting just outside the north entrance.*

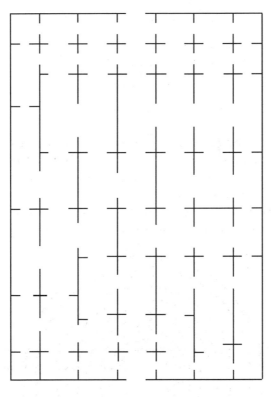

Figure 3.46

We can solve this puzzle by trying out several paths, tracing them with different-colored pencils, which, like Ariadne's string, leave traces. The paths leading nowhere can thus be identified until a correct one is wrested from the maze. Clearly, the exercise of accurate reckoning in solving these puzzles is critical. Here is the solution to the puzzle:

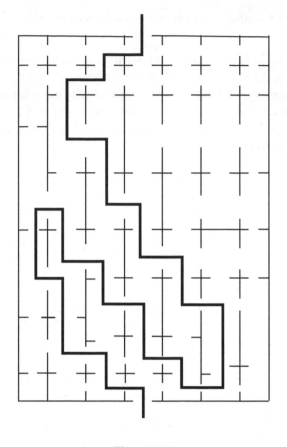

Figure 3.47

Today, all kinds of challenging mazes are being created to satisfy the many puzzle enthusiasts who cannot seem to get enough of them. Especially popular are hand-held *dexterity mazes,* in which objects—usually small balls—are manipulated inside a small maze structure. These are particularly popular with children.

The maze is also a symbolic element in many modern-day stories. Cortázar and Borges used the Cretan labyrinth as a metaphor for existence, and the same kind of fascination with the maze can be found, for instance, in *The Wizard of Oz,* by the American writer L. Frank Baum (1856–1919), which tells of a search for a wizard at the end of a maze-like yellow brick road. In the climax of Stephen King's (b. 1947) novel *The Shining,* the deranged principal character chases his son through a snow-covered garden maze, ax in hand. The boy escapes by creating a false set of footprints for his father to follow. The latter becomes hopelessly lost in the maze, collapsing into a state of catatonia. In Umberto Eco's (b. 1936) popular novel *The Name of the Rose* (1983), the central feature is a library constructed as an intricate labyrinth. The story tells of murders committed in a medieval monastery. A serial killer is loose among the monks. Two clerics from outside the monastery are called in to solve the murders. When first trapped within the library of the monastery, the two medieval sleuths are able to escape with the assistance of a thread, in an obvious allusion to the Cretan myth. At the center of the labyrinth they ultimately find the culprit—a blind monk who has poisoned the pages of Aristotle's book on humor. Fearful of the power of humor, the monk had taken it upon himself to eliminate all those who dared venture into the maze's center to discover the delights and dangers of laughter.

Why are labyrinths so appealing? Maybe the reason behind their appeal is itself a "hidden secret," to use Ahmes' apt phraseology once again, of the human mind. The universality and antiquity of labyrinths, along with the mystique that surrounds them, suggest that perhaps the structure of the mind is labyrinthine. If that is indeed true, then labyrinths would constitute reifications of the many twists and turns of human thought.

Order and Chaos

Whatever its motivation, the maze concept stands out as a counterpart to the Pythagorean belief in order, revealing the presence of a peculiar duality in the human soul—an inner struggle between order and chaos. Strangely, both seem to be required in geometry, Puzzleland, and life. In Chinese philosophy, this duality has been encapsulated by the notion of *yin* and *yang*—two opposing forces that are believed to combine in various proportions to produce all the different objects in the universe, as well as the different moods in human beings.

So petrified of chaos were the Pythagoreans that they considered their theory of order seriously undermined when, ironically, Pythagoras's own theorem revealed the existence of irrational numbers such as $\sqrt{2}$. This "chaotic" number stared them straight in the face each time they drew an isosceles right-angled triangle with equal sides of unit length. The length of its hypotenuse was the square root of the sum of $1^2 + 1^2$, or $\sqrt{2}$, a number that cannot be represented as the ratio of two integers, or as a finite or repeating decimal (it begins 1.14142136 . . .):

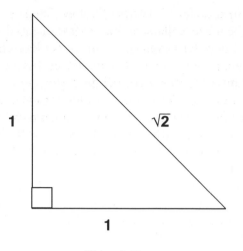

Figure 3.48

For the Pythagoreans, rational numbers had a "rightness" about them; irrational ones such as $\sqrt{2}$ did not. And yet there they were, defying logic and sense and challenging the system of order that the Pythagoreans so strongly desired to establish. So disturbed were they that, as some stories have it, they "suppressed their knowledge of the irrationality of $\sqrt{2}$, and went to the length of killing one of their own colleagues for having committed the sin of letting the nasty information reach an outsider" (Ogilvy 1956: 15). The colleague is suspected to have been Hipassus of Metapontum (Bunt, Jones, and Bedient 1976: 86; Aczel 2000: 19).

Incredibly, even in the domain of chaos, the human mind finds ways of making sense of things. *Chaos theory,* founded by the French mathematician Henri Poincaré (1854–1912), has shown, in fact, that there are patterns even in random events (Gleick 1987). In the early 1960s, simplified computer models demonstrated that there was a hidden structure in the seemingly chaotic patterns of weather. When these were plotted in three dimensions they revealed a butterfly-shaped fractal set of points. Similarly, leaves, coastlines, mounds, and other seemingly random forms produced by Nature reveal hidden fractal patterns when examined closely.

All this constitutes a truly profound paradox. Why is there order in chaos? Perhaps the ancient myths provide, after all, the only plausible response to this question. According to the *Theogony* of the Greek poet Hesiod (eighth century B.C.), Chaos generated Earth, from which arose the starry, cloud-filled Heaven. In a later myth, Chaos was portrayed as the formless matter from which the Cosmos, or harmonious Order, was created. In both versions, it is obvious that the ancients felt deeply that order arose out of chaos. This feeling continues to reverberate in the contemporary human imagination. For some truly mysterious reason, our mind requires that there be order within apparent disorder. The search for order has been the central motivating force of human history. And it will continue to be so as long as humans are around.

4 Puzzling Logic

DEDUCTIONS, PARADOXES, AND OTHER FORMS OF MIND PLAY

The paradox is really the pathos of intellectual life and
just as only great souls are exposed to passions it is only
the great thinker who is exposed to what I call paradoxes,
which are nothing else than grandiose thoughts in embryo.
—Søren Kierkegaard (1813–1855)

In the fifth century B.C. heated debates broke out frequently in Greece over one of the greatest philosophical puzzles of all time—how human beings think. Prominent in the debates were the philosopher Parmenides (born c. 515 B.C.) and his disciple Zeno of Elea (c. 489–c. 435 B.C.). In Zeno's time, leading scholars had based mathematics on the method of *deductive logic.* Zeno challenged this practice with a series of clever arguments. They were designated *paradoxes* (meaning literally "conflicting with expectation") by the people who witnessed the fascinating debates (Salmon 1970).

In one of his paradoxes, Zeno argued that a runner would never be able to reach a finish line. He argued his case as follows. The runner must first traverse half the distance to the finish line. Then, from that position, the runner would face a new, but similar, task—he must traverse half of the remaining distance between himself and the finish line. But thereupon, the runner would face a new, but again similar, task—he must once more cover half of the new remaining distance between himself and the finish line. Although the successive half distances between himself and the finish line would become increasingly (indeed infinitesimally) small, the wily Zeno concluded that the runner would come very close to the finish line, but would never cross it. The successive distances that the runner must cover form an infinite geometric sequence, each term of which is half the one before: {½, ¼, ⅛, ¹⁄₁₆, . . .}. The sum of the terms in this sequence—{½ + ¼ + ⅛ + ¹⁄₁₆ + . . .}—will never reach 1, the whole distance to be covered. Below is a visual portrayal of Zeno's runner paradox:

> *The runner will never reach the finish line because in order to do so, he must first traverse ½ of the distance to the finish line; and then half of that, i.e., ½ of ½ = ¼; and then half of that, i.e., ½ of ¼ = ⅛; and so on, ad infinitum. Hence, the runner can never reach the finish line:*

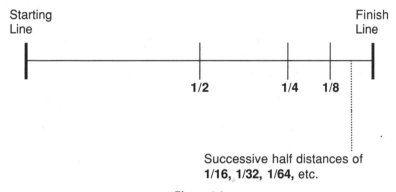

Figure 4.1

With such shrewdly contrived arguments, Zeno called into question the whole deductive-analytical edifice of Greek mathematics and philosophy. As Devlin (1998a: 101) aptly puts it, "Zeno's puzzles presented challenges to the attempts of the day to provide analytic explanations of space, time, and motion—challenges that the Greeks themselves were not able to meet."

The Sophists—a group of traveling teachers who became famous throughout Greece toward the end of the fifth century B.C.—sided with Zeno, arguing that the very existence of paradoxes meant that it was impossible to deduce anything worthwhile, contrary to what philosophers such as Thales and Socrates believed. Aristotle, on the other hand, dismissed Zeno's paradoxes as exercises in specious reasoning. The central characteristic of human thinking, Aristotle insisted, was its ability to deduce things from given facts. He then proceeded to give the deductive method a formal structure, which he designated *syllogistic:*

Major premise:	All mammals are warm-blooded.
Minor premise:	Whales are mammals.
Conclusion:	Therefore whales are warm-blooded.

The major premise states that a category has (or does not have) a certain characteristic, and the minor premise states that a certain thing is a member of the given category. The conclusion then affirms (or denies) that the thing has that characteristic. As clever as they were, Aristotle stated, Zeno's paradoxes were inconsequential, since they did not impugn the validity of the syllogism, which is based on the logic of common sense. But Aristotle's attempt to demolish Zeno's clever arguments did not eradicate them from the history of logic and mathematics. On the contrary, they became important factors in the development of those two disciplines, and are still being debated heatedly today. As Kasner and Newman (1940: 39) perceptively state, the "history of mathematics, in fact, recounts a poetic vindication of Zeno's stand."

Aristotelian logic was elaborated further in the nineteenth and early twentieth centuries by George Boole (1815–1864), Augustus De Morgan (1806–1871), Gottlob Frege (1848–1925), Giuseppe Peano (1858–1932), Bertrand Russell (1872–1970), and Alfred North Whitehead (1861–1947). But despite their valiant attempts to make it, once and for all, the cornerstone of the edifice of mathematics, they failed to do so, mainly because of the work of two modern-day descendants of Zeno (as they can be freely characterized here), Georg Cantor (1845–1918) and Kurt Gödel (1906–1978), who came forward to put a damper on their well-intentioned efforts.

In one of his demonstrations, Cantor defied Aristotle's "logic of common sense" by proving that the set of even integers ($\{2, 4, 6, 8, \dots\}$), which is a subset of the integers, is the same size as the entire set of integers, even and odd ($\{1, 2, 3, 4, \dots\}$)! He did this rather matter-of-factly, by putting the even integers in a one-to-one correspondence with the complete set of integers. Cantor showed that there are no "leftovers," no matter how long one continues the one-to-one matching. In other words, every even integer will be matched with exactly one integer, and vice versa [see figure 4.2].

Cantor's demonstration put a fly in the logical ointment of common sense, so to speak, because it showed that the infinite set of even integers contains the same number of objects as the set of which it is a part, the infinite set of all the integers. How can this be? The reason is, of course, that any number n can be rendered even by simply multiplying it by 2 (yielding $2n$). For every n there will always be a $2n$. Still, the demon-

1	2	3	4	5	6	7	8	*and so on, ad infinitum*
↕	↕	↕	↕	↕	↕	↕	↕	
2	4	6	8	10	12	14	16	*and so on, ad infinitum*

Figure 4.2

stration, as simple as it is, was earth-shattering when it was first made public by Cantor. Its aftershocks are still being felt today in mathematical circles.

Kurt Gödel put another fly in the Aristotelian ointment in 1931. Before Gödel, it was taken for granted that every proposition within a logical system could be either proved or disproved within that system. But Gödel startled the mathematical world by showing that this was not the case! He showed that a logical system invariably contains a proposition that is true but unprovable within it. For illustrative purposes, Gödel's ingenious demonstration can be paraphrased as follows:

> Consider a mathematical system T that is both *correct*—in the sense that no false sentence is provable in it—and contains a sentence S that asserts its own unprovability in the system. S can be formulated simply as "I am not provable in system T." What is the truth status of the sentence S? If it is false, then its opposite is true, which means that S is provable in system T, contrary to our assumption that no false sentence is provable in the system. Therefore we conclude that S must be true, from which it follows that S is unprovable in T, as S asserts.

Responding to Russell and Whitehead's three-volume work on symbolic logic, *Principia Mathematica* (1910–1913), Gödel proved, in effect, that any consistent (correct) mathematical system is incomplete, because formulae (sentences, propositions, etc.) can be constructed that can neither be proved nor disproved within the system. His demonstration thus shattered, once and for all, the dream of building a mathematical edifice of truths, all resting on the cornerstone of Aristotelian logic.

Incidentally, the American puzzlist Raymond Smullyan (1997: 152) provides a clever puzzle version of Gödel's proof as follows:

> *Let us define a logician to be accurate if everything he can prove is true; he never proves anything false.*
> *One day, an accurate logician visited the Island of Knights and Knaves, in which each inhabitant is either a knight or a knave, and knights make only true statements and knaves make only false ones. The logician met a native who made a statement from which it fol-*

lows that the native must be a knight, but the logician can never prove that he is!

What statement would work?

The "Gödelian" statement that would work is: *You cannot prove that I am a knight*. Why? Assume that the native is a mendacious knave. Then, his statement would be false. But in that case, the logician would, in fact, be able to prove that the native was a knight—contrary to what the native says. But the native is not a knight—since that is, in fact, our initial assumption—no matter what the logician is capable of proving. And, as a consequence, the logician would be proving something false—contrary to the condition that he is "accurate." Therefore, the only conclusion we can safely draw from this line of reasoning is that, contrary to our initial assumption, the native must be a knight. This means that his statement is true. But if his statement is true, then the logician cannot prove that the native is a knight—the statement declares as much. So, even though the native is a knight, the logician will never be able to prove it!

As Smullyan's clever puzzle insinuates—as do, indeed, all of his superb puzzles—paradoxes are, fundamentally, products of the puzzle instinct. Like riddles, they expose human logic as Janus-faced—both useful and deceptive. Paradoxes are found in a domain of Puzzleland that is inhabited by the Carrollian characters of Tweedledum and Tweedledee, who were fascinated by the incongruities of logical thinking itself. Carroll, incidentally, was the first puzzlist to show a keen interest in the formal structure of logic, producing the first modern genre of *logic puzzles* (Carroll 1958a). Since then, collections of logic puzzles have become highly popular across the world. For the sake of convenience, they can be divided into four main categories: (1) *deduction puzzles*, i.e., puzzles that require the solver to draw conclusions from various statements; (2) *truth puzzles*, i.e., puzzles that require the solver to assess the consistency of a certain set of statements; (3) *deception puzzles*, i.e., puzzles that lead the solver astray or into logical traps; and (4) *paradoxes* à la Zeno.

Deduction Puzzles

Despite Zeno's clever arguments, Aristotle's assertion that deductive logic allows us to make sense of the categories and relations of everyday experience remains largely true. In our three-dimensional world, if it is true, for instance, that staircase A is higher than staircase B, and that staircase C is higher than A, then it can be concluded without any shadow of a doubt that staircase C is higher than staircase B. But, then, Zeno would warn us that this apparently commonsensical logical deduction is not always so. As we saw in the previous chapter, it would not hold, for instance, in a Möbius-type space where the concepts of "higher" and "lower" simply do not apply. But this counterperspective need not concern us at this point.

Puzzles based on deductive logic bring out the Aristotelian perspective rather forcefully. These fall into four main categories: (1) those that require the solver to draw conclusions from a series of statements; (2) those based on set theory; (3) those that play on certain relations, such as kinship; and (4) those that require the solver to use pure inferential reasoning. The first type can be called simply *deductions,* the second *set-theoretical puzzles,* the third *relational puzzles,* and the fourth *inferential puzzles.*

Deduction puzzles were first proposed as a distinct genre by Henry Dudeney (1958a). The following is a paraphrase of one of his puzzles; one similar to it is included in all standard collections of logic puzzles:

> *In a certain company, Bob, Janet, and Shirley hold the positions of director, engineer, and accountant, but not necessarily in that order. The accountant, who is an only child, earns the least. Shirley, who is married to Bob's brother, earns more than the engineer. What position does each person fill?*

To solve this puzzle systematically, it is useful to have a cell chart, such as the one below. It will allow us to keep track of the various deductions we make in the course of the solution.

	Director	Engineer	Accountant	← positions
Bob				
Janet				
Shirley				

↑

persons

Figure 4.3

If we conclude, for instance, that one of the three people cannot be the director, then we can place an × in the cell opposite the person's name under the column headed *Director.* If we deduce that one of the three is the engineer, then we can put a different mark, •, opposite that person's name under the column headed *Engineer,* eliminating the remaining cells in the column (since there can be only one engineer) with ×s. The solution is complete when we have placed exactly one • in each row and column in a way logically consistent with the puzzle's constraints.

Dudeney's puzzle puts on display the power of deductive reasoning to flesh out patterns in given facts. It shows, in other words, the value of Aristotelian logic in everyday life. We are told that the accountant is an only child, but that Bob has a brother. So, clearly, Bob is not the accountant, and we can place an × in the cell opposite his name under the column headed *Accountant.* We are also told that the accountant earns the least of the three, but that Shirley earns more than the engineer does. From these two statements, we can deduce two things about Shirley: (1) that she is not the accountant (who earns the least, while she earns more than someone else does); and (2) that she is not the engineer (since she earns more than that person does). To keep track of these deduced facts, we enter ×s in their appropriate cells, eliminating *Accountant* as a possibility for *Bob* and both *Accountant* and *Engineer* as possibilities for *Shirley.* The chart will then look like this:

	Director	Engineer	Accountant
Bob			X
Janet			
Shirley		X	X

Figure 4.4

The chart itself now reveals that the only cell left under *Accountant* is opposite *Janet.* Therefore, by the process of elimination, Janet is the accountant. We show this by putting the mark • opposite her name in the appropriate cell, and eliminating all other possibilities for *Janet* with ×s (for she can hold only one of the stated positions):

	Director	Engineer	Accountant
Bob			X
Janet	X	X	●
Shirley		X	X

Figure 4.5

The chart then shows that the only cell left under *Engineer* is opposite *Bob*. So we can put the • mark opposite *Bob* under *Engineer*, eliminating all other possibilities for *Bob* with ×s. One last look at the chart will then show us that the only cell left opposite *Shirley* is under *Director*:

	Director	Engineer	Accountant
Bob	X	•	X
Janet	X	X	•
Shirley	•	X	X

Figure 4.6

We have discovered, with simple deductive reasoning, that Bob is the engineer, Janet the accountant, and Shirley the director. The fundamental thing to notice about the cell chart is that it itself will, at certain points, guide the search for a solution, since it reveals connections on its own. This underlines the usefulness of visual representation in problem-solving. Diagrams and charts literally allow us to "figure out" what is going on. Indeed, all theories of logic are, *ipso facto,* diagrammatic artifacts, i.e., arrangements of symbols on a page, showing the connections that inhere among them (called the *syntax* of the logical system). Success at solving virtually any logic puzzle is dependent upon knowing its syntax, which, in turn, depends on knowing how to represent the problem diagrammatically. This suggests, incidentally, that even the most abstract forms of reasoning, such as the deductive form, are based on mental imagery.

Consider, as another example, the following classic in this genre, which is a version of another puzzle devised by Henry Dudeney:

> The president, publicist, and CEO of a certain company are Ms. Sainz, Ms. Joaquín, and Ms. Robinson, but not necessarily in that order. Working for the same company are three male custodians who have the same surnames (but who are not related or married to the three women)—Mr. Sainz, Mr. Joaquín, and Mr. Robinson.
>
> 1. Mr. Robinson lives in Detroit.
> 2. The president lives halfway between Chicago and Detroit.
> 3. Mr. Joaquín has one child.
> 4. Ms. Sainz regularly beats the publicist at chess.

5. *The president's next-door neighbor, one of the custodians, has exactly three times as many children as one of the other custodians.*

6. *The custodian who lives in Chicago has the same surname as the president.*

What positions do Ms. Sainz, Ms. Joaquín, and Ms. Robinson fill?

Again, we start by setting up a cell chart relating two sets—the surnames of the female employees (*Sainz, Joaquín, Robinson*) and their positions (*president, publicist, CEO*):

	President	Publicist	CEO
Sainz			
Joaquín			
Robinson			

Figure 4.7

The president's surname can be determined by finding out the surname of the custodian who lives in Chicago, since that custodian has the same surname as the president (statement 6). Statement 1 tells us that Mr. Robinson lives in Detroit. So the president's surname is certainly not Robinson. We can also deduce that the custodian living next to the president is not the one who lives in Chicago (i.e., the one with the same surname), because statement 2 tells us that the president lives halfway between Chicago and Detroit, not in Chicago. Also, we are told that the president's next-door neighbor has three times as many children as another custodian (statement 5). Since Mr. Joaquín has one child (statement 3), it is obvious that he cannot be her neighbor. So Mr. Joaquín is definitely not the president's next-door neighbor. Her neighbor is, therefore, Mr. Sainz, who thus also lives halfway between Chicago and Detroit. We can now eliminate Sainz as a possible surname for the president.

The above deductions allow us to fill in our chart as follows:

	President	Publicist	CEO
Sainz	X		
Joaquín			
Robinson	X		

Figure 4.8

The chart now reveals that the president's surname is Joaquín, since the only cell left under president is opposite Joaquín:

	President	Publicist	CEO
Sainz	X		
Joaquín	●	X	X
Robinson	X		

Figure 4.9

Statement 4 tells us that Ms. Sainz beats the publicist regularly at chess. So Ms. Sainz is definitely not the publicist:

	President	Publicist	CEO
Sainz	X	X	
Joaquín	●	X	X
Robinson	X		

Figure 4.10

As we can now see from the chart, Ms. Sainz is the CEO—since the only cell opposite Sainz is under CEO. By the process of elimination, Ms. Robinson is the publicist:

	President	Publicist	CEO
Sainz	X	X	●
Joaquín	●	X	X
Robinson	X	●	X

Figure 4.11

Incidentally, although Dudeney was the inventor of this type of puzzle, its widespread popularity as a genre was due to the British puzzlist Hubert Phillips (1933, 1934, 1936, 1937), who concocted many ingenious deduction puzzles in the 1930s under the pseudonyms of "Caliban" for the *New Statesman* and "Dogberry" for the *News Chronicle.*

The great fictional detectives have all excelled at deductive reasoning. The first such character, Edgar Allan Poe's Auguste Dupin, appeared in April 1841, when *Graham's Magazine* published Poe's classic mystery story "The Murders in the Rue Morgue." Following Poe's successful prototype, shortly thereafter Sir Arthur Conan Doyle (1859–1930) created Sherlock Holmes; G. K. Chesterton (1874–1936) Father Brown, a priest-detective; and Agatha Christie (1891–1976) Hercule Poirot, a dapper detective who, like Holmes, would explain his deductions to the reader. By solving deduction puzzles, one is engaging in the same kind of mental detective work, drawing inferences from the available evidence and reaching the only possible conclusion from the given facts.

Set-theoretical puzzles were invented by Lewis Carroll, foreshadowing the more serious work of Cantor and other set-theory mathematicians. Here is one of his puzzles in this genre (Carroll 1958b: 12):

> *What can you conclude from the following statements? (1) Babies are illogical. (2) Nobody is despised who can manage a crocodile. (3) Illogical persons are despised.*

We start our solution by identifying the relevant sets: $D =$ persons who are despised; $not\text{-}D =$ persons who are not despised; $I =$ illogical persons; $B =$ babies; $C =$ persons who can manage a crocodile. The sets D and $not\text{-}D$ hold no common elements, because persons belong to one or the other, but not to both. So we draw a square to represent D, with the space around it representing $not\text{-}D$; a person is either in the square and despised, or outside it and not despised. Now, I is a subset of D, because statement (3) asserts that illogical persons are despised. This can be shown with a circle inscribed inside the D square. Statement (1) informs us that babies are illogical. This means that B is itself a subset of I. This can be shown with a smaller B circle inscribed within the larger I circle inside the D square. Finally, the set C, which consists of persons who can manage a crocodile, is a subset of $not\text{-}D$ (= persons who are not despised), as statement (2) tells us. C can thus be represented by a circle inscribed in the space surrounding the D square [see figure 4.12]:

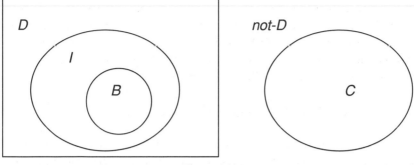

Figure 4.12

Now, from the above diagram, it can be seen that no baby exists who can manage a crocodile. And that is the solution to Carroll's puzzle.

Since Carroll's time, many puzzlists have created ingenious set-theoretic puzzles. The following one is typical. Its solution is left as an exercise for the reader (see solution 4.1):

> *Jill and Franklin are grade 10 teachers in a high school where many sports-minded students are enrolled. Jill teaches math, Franklin history. The other day they were discussing their students at lunch. The topic of sports competition came up.*
>
> *"Did you know that our students have become highly competitive in hockey and soccer?" Jill remarked.*
>
> *"Really?" answered Franklin.*
>
> *"Moreover," continued Jill, "every one of the grade 10 students is involved in one or both of these sports. And, I should add, this is the largest grade 10 class we have had in the last five years."*
>
> *"Do you know how many students there are altogether in grade 10?" inquired Franklin.*
>
> *"Yes, and I will tell you," Jill replied playfully, "but only in the form of a set theory puzzle."*
>
> *"As always, ask a math teacher a question, and you will get a puzzle," replied Franklin dryly. "OK, shoot!"*
>
> *"There are 180 students who are involved in hockey. Seventy percent of the students play soccer, and 10 percent play both hockey and soccer," Jill declared.*
>
> *The question is: How many students were there altogether in grade 10?*

Set theory is, in a basic sense, diagram theory. It was George Boole and

John Venn (1834–1923) who established the standard practice of using diagrams as the representational basis for discussing logical relations.

Henry Dudeney was also an ingenious creator of *relational puzzles*. These require the solver to identify a relation of some kind, usually kinship. Here is one of Dudeney's classics in this genre:

> A boy is looking at a photo: "Brothers and sisters have I none, but this man's son is my father's son." Who is the person in the picture?

The boy is an only child, because he says that he has no brothers or sisters. So, when he refers to "my father's son" he is referring to himself. Now, we can use this equivalence relation—namely, *my father's son = I*—to make sense of the key statement in the puzzle—*this man's son is my father's son*—as follows:

(1) *this man's son = my father's son*
(2) *my father's son = I*

Therefore, substituting (2) into (1), we get

(3) *I = this man's son*

The boy is, thus, looking at a portrait of his father.

To the best of my knowledge, *inferential puzzles* were probably invented by Hubert Phillips (mentioned on p. 122). These involve setting up hypothetical situations and then reasoning about their logical implications. Here's a version of one of Phillips's puzzles in this genre:

> Before they are blindfolded, three women are told that each one will have either a red or a blue cross painted on her forehead. When the blindfolds are removed, each is supposed to raise her hand if she sees a red cross and to drop her hand when she figures out the color of her own cross. Now, here's what actually happens. The three women are blindfolded and a red cross is drawn on each of their foreheads. The blindfolds are removed. After looking at each other, the three women raise their hands simultaneously. After a short time, one of the women lowers her hand and says, "My cross is red." How did she figure it out?

Solving this puzzle requires the same kind of mental projection that was used to solve the puzzle of the fast and slow watches in chapter 1. It hinges on projecting oneself into the minds of the characters of the puzzle, so as to envision what each one thinks as she looks at the others' blindfolds. We can start, for the sake of clarity, by calling the three women A, B, and C. We'll assume that A is the one who figured out the color of the cross on her head. (A is mentally adroit and thus quicker than the other two.) Let us now enter A's mind. She looks at B and C and sees that they both have red crosses. So, naturally, she will put up her hand as she has been instructed to do. Similarly, B also sees two red crosses. So she too raises her hand. C likewise sees two red crosses, and of course she too will raise her hand. At that point, A reasons as follows:

> *Let me assume that I have a blue cross on my forehead. If that is so, then one of the other two, say B, would know that she doesn't have a blue cross, because otherwise C, seeing two blue crosses—mine and B's—would not have put up her hand. But this has not occurred. So B and C cannot determine their colors. This means that I too have a red cross.*

Truth Puzzles

The main requirement of deductive logic is *consistency*—the inability of a set of statements to yield contradictory conclusions. Puzzles that play on consistency directly are called *truth puzzles;* they were also invented by Hubert Phillips. They fall into two main categories: (1) those containing statements made by individuals (suspects in a murder or robbery, for example), some of which are true and others false; (2) those that require the solver to identify to what tribe or group an individual or individuals belong on the basis of certain statements that are made.

The following is an example of type (1). A detailed explanation of its solution is given below because all puzzles of this type are solved in exactly the same way.

> *Mack "the Knife," the notorious tough guy, was found murdered one night in an alley behind the nightclub he usually frequented. The police brought in three suspects the morning after. That afternoon a police investigator interrogated the three men. They made the following statements.*

Bud: 1. *I didn't kill Mack.*
     2. *Jack is not my friend.*
     3. *I knew Mack.*

Jack: 1. *I didn't kill Mack.*
     2. *Bud and Tug are friends of mine.*
     3. *Bud didn't kill Mack.*

Tug: 1. *I didn't kill Mack.*
      2. *Bud lied when he said that Jack was not his friend.*
      3. *I don't know who killed Mack.*

Only one of the three is guilty, and only one of each man's statements is false. Who killed Mack the Knife?

As with deductions, it would be a mind-boggling task to keep track of all the possible true-false arrangements that this puzzle generates without some visual device to help us keep track of them. So the first thing to do is to set up a *truth chart* in which we can display the statements of all three men and record their truth values as we make our inferences and deductions. The solution to this puzzle is complete when we have placed exactly one F (for a false statement) and two Ts (for true statements) opposite each man's three statements in a consistent way—because the puzzle informs us that only one of each man's statements is false (making two of them necessarily true).

	Statement	Truth Value
Bud	1. I didn't kill Mack.	1.
	2. Jack is not my friend.	2.
	3. I knew Mack.	3.
Jack	1. I didn't kill Mack.	1.
	2. Bud and Tug are friends of mine.	2.
	3. Bud didn't kill Mack.	3.
Tug	1. I didn't kill Mack.	1.
	2. Bud lied when he said that Jack was not his friend.	2.
	3. I don't know who killed Mack.	3.

Figure 4.13

The first thing to note is that the initial statements are the same— *I didn't kill Mack.* Since we are told that one of the three is the murderer, then one of these statements is necessarily false. Let us assume that Bud's first statement is the false one. This would make him the murderer. Then the first statements of the other two suspects are necessarily true (since there can be only one murderer). So we can go ahead and write an F opposite Bud's first statement and Ts opposite the corresponding statements of the other two suspects. Having done this, we can follow the logical implications of our assumption in order to see where it leads. We start out on our journey by adding Ts opposite Bud's second and third statements, since we know that only one of his three statements is false:

	Statement	Truth Value
Bud	1. I didn't kill Mack.	1. F
	2. Jack is not my friend.	2. T
	3. I knew Mack.	3. T
Jack	1. I didn't kill Mack.	1. T
	2. Bud and Tug are friends of mine.	2.
	3. Bud didn't kill Mack.	3.
Tug	1. I didn't kill Mack.	1. T
	2. Bud lied when he said that Jack was not his friend.	2.
	3. I don't know who killed Mack.	3.

Figure 4.14

We note that Bud's second statement—*Jack is not my friend*—now has a truth value of T. Consider Jack's second statement—*Bud and Tug are friends of mine.* Is it true or false? It contradicts Bud's second statement, which we have marked true. So Jack's second statement is apparently false. Similarly, consider Tug's second statement—*Bud lied when he said that Jack was not his friend.* It too is false, because it too contradicts Bud's second statement. The liar is, therefore, Tug. We can now write these two new Fs in the chart—one opposite Jack's second statement and one opposite Tug's second statement. It is now an easy task to complete the chart. We know that each suspect made one false (F) and two true (T) statements. For both Jack and Tug, we have established one F and one T statement. So we can go ahead and add the remaining T to Jack's and Tug's statements—namely opposite their third statements:

	Statement	Truth Value
Bud	1. I didn't kill Mack.	1. F
	2. Jack is not my friend.	2. T
	3. I knew Mack.	3. T
Jack	1. I didn't kill Mack.	1. T
	2. Bud and Tug are friends of mine.	2. F
	3. Bud didn't kill Mack.	3. T
Tug	1. I didn't kill Mack.	1. T
	2. Bud lied when he said that Jack was not his friend.	2. F
	3. I don't know who killed Mack.	3. T

Figure 4.15

However, the assumption that Bud is the murderer leads to an inconsistency. Consider Jack's third statement—*Bud didn't kill Mack.*

According to the chart, it is a true statement. So, according to the chart, Bud is not the murderer. But the chart also says that he is, because his statement that he did not kill Mack is marked F. So, is he or isn't he the murderer? Obviously, we cannot tell from this particular arrangement of Fs and Ts. It is an arrangement that turns out to be logically inconsistent. What caused the inconsistency? It can only be the assumption with which we started our logical journey—namely that Bud's first statement is his false one. So it would appear that his first statement is definitely not his false one; it must be true. We can go ahead and record this sure finding in our chart, erasing all the other Fs and Ts from it:

	Statement	Truth Value
Bud	1. I didn't kill Mack.	1. T
	2. Jack is not my friend.	2.
	3. I knew Mack.	3.
Jack	1. I didn't kill Mack.	1.
	2. Bud and Tug are friends of mine.	2.
	3. Bud didn't kill Mack.	3.
Tug	1. I didn't kill Mack.	1.
	2. Bud lied when he said that Jack was not his friend.	2.
	3. I don't know who killed Mack.	3.

Figure 4.16

Since Bud's first statement—*I didn't kill Mack*—has now been established as true, we can see that Jack's third statement is also true— *Bud didn't kill Mack*—because it simply confirms what Bud said in his first statement. So we can safely put a T in the chart opposite Jack's third statement. Let us assume that Tug's first statement is his false one. This identifies Tug as the murderer. We can now insert Tug's two Ts opposite his second and third statements. We write these truth values in some different font or style, say in italics, so as to distinguish them from the two sure truth values we have established so far—namely, the T opposite Bud's first statement and the T opposite Jack's third statement [see figure 4.17].

Where does this lead? Consider Jack's first statement—*I didn't kill Mack*. Since we have assumed that Tug is the murderer, it is obvious that Jack did not kill Mack. His first statement simply confirms this. So it is true. We can go ahead and write an italic T (to indicate that it depends on our initial assumption) opposite Jack's first statement [see figure 4.18].

Looking at Jack's set of statements, we note that there are two Ts in

	Statement	Truth Value
Bud	1. I didn't kill Mack.	1. T
	2. Jack is not my friend.	2.
	3. I knew Mack.	3.
Jack	1. I didn't kill Mack.	1.
	2. Bud and Tug are friends of mine.	2.
	3. Bud didn't kill Mack.	3. T
Tug	1. I didn't kill Mack.	1. *F*
	2. Bud lied when he said that Jack was not his friend.	2. *T*
	3. I don't know who killed Mack.	3. *T*

Figure 4.17

	Statement	Truth Value
Bud	1. I didn't kill Mack.	1. T
	2. Jack is not my friend.	2.
	3. I knew Mack.	3.
Jack	1. I didn't kill Mack.	1. *T*
	2. Bud and Tug are friends of mine.	2.
	3. Bud didn't kill Mack.	3. T
Tug	1. I didn't kill Mack.	1. *F*
	2. Bud lied when he said that Jack was not his friend.	2. *T*
	3. I don't know who killed Mack.	3. *T*

Figure 4.18

it—one opposite his first statement and one opposite his third statement. This means that Jack's second statement must be false. Now, consider Bud's three statements. We know that his first one is true. His second statement—*Jack is not my friend*—is false, because Tug's second statement—*Bud lied when he said that Jack was not his friend*—is true. That means that when Bud said *Jack is not my friend,* he lied. So Bud's second statement must be false and, thus, his third statement is necessarily true. We can now complete the chart as shown [see figure 4.19].

Is this new arrangement of Fs and Ts logically consistent? Consider Jack's second statement—*Bud and Tug are friends of mine.* Since it is false, Bud is not one of Jack's friends. So Bud's second statement—*Jack is not my friend*—is in fact true. But, according to the chart, it has a truth value of F; it is false. The second arrangement also turns out to be logically inconsistent. What caused the inconsistency this time? It can only be the assumption with which we started our second logical journey—namely

	Statement	Truth Value
Bud	1. I didn't kill Mack. 2. Jack is not my friend. 3. I knew Mack.	1. T 2. *F* 3. *T*
Jack	1. I didn't kill Mack. 2. Bud and Tug are friends of mine. 3. Bud didn't kill Mack.	1. *T* 2. *F* 3. T
Tug	1. I didn't kill Mack. 2. Bud lied when he said that Jack was not his friend. 3. I don't know who killed Mack.	1. *F* 2. *T* 3. *T*

Figure 4.19

that Tug's first statement is his false one. As it turns out, it is definitely not his false one. It is therefore necessarily one of his two true ones. So, first we erase all the italicized Fs and Ts from the chart, leaving in it only the two Ts that were established during our first logical journey. We can now record a new truth value of T opposite Tug's first statement:

	Statement	Truth Value
Bud	1. I didn't kill Mack. 2. Jack is not my friend. 3. I knew Mack.	1. T 2. 3.
Jack	1. I didn't kill Mack. 2. Bud and Tug are friends of mine. 3. Bud didn't kill Mack.	1. 2. 3. T
Tug	1. I didn't kill Mack. 2. Bud lied when he said that Jack was not his friend. 3. I don't know who killed Mack.	1. T 2. 3.

Figure 4.20

We have now identified the murderer. It is, of course, Jack, whose first statement—*I didn't kill Mack*—has been shown by the process of elimination to be the false one. Nevertheless, it is always wise to complete the chart by deductive reasoning, just to make sure that no inconsistencies arise from this finding (see solution 4.2).

Truth puzzles are sometimes called *mystery puzzles,* because solving them is akin to solving a crime mystery. In setting up a hypothesis and following it through to see where it leads, the solver is using the same kind of reasoning process as Sherlock Holmes or Hercule Poirot. Indeed, many truth puzzles are cast in a detective narrative form for this very reason.

The second type of truth puzzle involves determining which individuals belong to which tribe, clan, group, etc., on the basis of certain statements. The following is a version of one of Phillips's classic puzzles in this genre:

> The people of an island culture belong to two tribes—the Bungu and the Mungu. Since members of both tribes look and dress alike, and since they speak the same language, they are virtually indistinguishable. It is known, however, that the members of the Bungu tribe always tell the truth, whereas the members of the Mungu tribe always lie. The anthropologist who became interested in their fascinating social system, Dr. Mary Truthspin, recently came across three men on the island.
>
> "To which tribe do you belong?" Dr. Truthspin asked the first.
> "Dutu luneh," he replied in his native language.
> "What did he say?" she asked the second and third men, both of whom had learned to speak some English.
> "He said that he is a Bungu," said the second.
> "No, he said that he is a Mungu," said the third.
> Can you figure out to which tribes the second and third men belonged?

The method used in solving this genre of puzzle is to zero in on a specific statement in order to test its truth or consistency, or else, as in this case, to unravel its actual content. The key, therefore, lies in translating *Dutu luneh* into English. Assume that the first individual belonged to the truth-telling Bungu tribe. His answer to the anthropologist's question *To which tribe do you belong?* would have been, of course, that he belonged to the Bungu tribe, because as a Bungu he would not lie. So, in this hypothetical scenario, *Dutu luneh* translates as *I belong to the Bungu tribe.*

Now, assume the opposite, namely that the first man belonged to the mendacious Mungu tribe. His answer to Truthspin's question would have been a lie. He certainly would not have admitted belonging to the Mungu tribe. Instead, he would have lied and said that he belonged to the other tribe, the Bungu. Once again, in this second hypothetical scenario *Dutu luneh* translates as *I belong to the Bungu tribe.*

In sum, no matter to which tribe the first man really belonged, the anthropologist would have gotten the same answer from him. Now, consider the responses given by the other two men to Truthspin's follow-up question—*What did he say?* We start with the second individual's response:

> *Dr. Truthspin: What did he say?*
> *Second Man: He said that he is a Bungu.*

As we have just discovered, the first man did indeed say that he was a Bungu. So the second told the truth. That means that he himself was a member of the Bungu tribe.

Finally, consider the response given by the third man:

> *Dr. Truthspin: What did he say?*
> *Third Man: He said that he is a Mungu.*

As we know, the first man said that he was a Bungu. So the third clearly lied. This means, of course, that he was a member of the deceitful Mungu tribe. It is, of course, not possible to determine to which tribe the first man belonged.

In the 1970s, Raymond Smullyan became the undisputed master of this type of logic puzzle. Here is a paraphrased example of his delightful art (Smullyan 1997: 48–49):

> *"News has reached me, O Auspicious King, of a curious town in which every inhabitant is either a Mino or an Amino."*
> *"Oh my goodness, what are they?" asked the king.*
> *"The Minos are worshippers of a good god; whereas the Aminos worship an evil god. The Minos always tell the truth—they never lie. The Aminos never tell the truth—they always lie. All members of one family are of the same type. Thus given any pair of brothers, they are either both Minos or both Aminos. Now, I heard a story of two brothers, Bahman and Perviz, who were once asked if they were married. They gave the following replies:*
>
> *Bahman: We are both married.*
> *Perviz: I am not married.*
>
> *Is Bahman married or not? And what about Perviz?"*

We know that the two statements are either both true or both false, since the brothers are of the same religion. They obviously cannot both be true, since they contradict each other. So they are both false. Because Perviz's statement is false, we can take from it the opposite of what he said: Perviz is married. Bahman's statement is also false; they are not both married. Since we have established that Perviz is married, we conclude that Bahman is not.

Smullyan follows this puzzle up with a more challenging one:

"According to another version of the story, O Auspicious King, Bahman didn't say that they were both married; instead he said, 'We are both married or both unmarried.' If that version is correct, then what can be deduced about Bahman and what can be deduced about Perviz?"

It is useful to write out the new pair of statements, for the sake of clarity:

> *Bahman: We are both married or both unmarried.*
> *Perviz: I am not married.*

If both statements are true, then both brothers are unmarried. If both statements are false, then Perviz is married, contrary to what he says. Now, since Bahman's statement is false and Perviz is married, Bahman is unmarried. No other conclusion can be drawn from Bahman's mendacious statement. In either case, therefore, Bahman is unmarried, and we can conclude that this is true.

Phillips characterized the intellectual pleasure that such puzzles provide as an *aesthetics of mind.* He put it as follows (1934: iv): "The invention of such exercises, and the solving of them, both give great pleasure, since their construction can involve—and in my view should have reference to—principles of artistry which embody an aesthetic of their own." In fact, Phillips's characterization applies to the pleasure, or *Aha!* effect, that ensues from solving virtually any kind of puzzle. As will be discussed in the final chapter, a puzzle is indeed a small work of art that stimulates curiosity and provides a kind of aesthetic pleasure all its own.

Deception Puzzles

The makers of riddles capitalized on the Janus-faced nature of language, as we saw in chapter 2. The makers of what are called *deception puzzles* have exploited the ambiguities built into language in parallel ways. Here is a typical example of a deception puzzle:

> *How many cubic inches of dirt are there in a hole that is one foot deep, two feet wide, and six feet long?*

This puzzle is written in a style that resembles an ordinary problem in arithmetic. But there is a linguistic trap in it—a hole doesn't contain dirt; it is what is left after the dirt is removed. Because of the puzzle's style, unsuspecting solvers are duped into calculating the number of cubic inches of dirt taken out from the ground. But the puzzle does not ask us to do that.

Here are various other deception puzzles, containing similar kinds of traps (see solutions 4.3 to 4.8):

> *If you saw three shadows on three fence posts, one painted white, one painted blue, and one painted red, which shadow would be the darkest?* (solution 4.3)

> *How many times can you subtract the number one from the number twenty-five?* (solution 4.4)

> *A train leaves New York for Chicago traveling at the rate of 100 miles per hour. Another train leaves Chicago for New York an hour later, traveling at the rate of 75 miles per hour. When the two trains meet, which one is nearer New York?* (solution 4.5)

> *If 3 cats kill 3 rats in 3 minutes, how long will it take 100 cats to kill 100 rats?* (solution 4.6)

> *If it takes 3¾ minutes to boil one egg, how long will it take to boil six eggs?* (solution 4.7)

> *A farmer had seven daughters, and they each had a brother. How many children did he have?* (solution 4.8)

As these puzzles show, the trap is hidden in the meaning of certain words, the pseudo-mathematical style of the problem, or the presentation of the facts. The following classic deception puzzle never fails to confound solvers who come across it for the first time:

> *Three women decide to go on a holiday to Las Vegas. They share a room at a hotel which is charging 1920s rates as a promotional gimmick. The women are charged only $10 each, or $30 in all. After going through his guest list, the manager discovers that he has made a mistake and has actually overcharged the three vacationers. The room the three are in costs only $25. So he gives a bellhop $5 to return to them. The sneaky bellhop knows that he cannot divide $5 into three equal amounts. Therefore, he pockets $2 for himself and returns only $1 to each woman.*
>
> *Now, here's the conundrum. Each woman paid $10 originally and got back $1. So, in fact, each woman paid $9 for the room. The three of them together thus paid $9 × 3, or $27 in total. If we add this amount to the $2 that the bellhop dishonestly pocketed, we get a total of $29. Yet the women paid out $30 originally! Where is the other dollar?*

The trap in this puzzle is not to be found in any single word, but in the way in which the numerical facts are laid out. Here is how the facts should have been laid out in order to avoid the apparent discrepancy. Originally, the women paid out $30 for the room. That is how much money was in the hands of the hotel manager when he realized that he had overcharged them. He kept $25 of the $30 and gave $5 to the bellhop to return to the women. Each woman got back $1. This means that each one paid $9 for a room. Thus, altogether the three women spent $27. Of this money, the hotel got $25 and the other $2 was pilfered by the devious bellhop. So there is no missing dollar.

Another way to explain this puzzle is as follows. Again, we start by noting that the women paid out $30 for the room. Of this money, the manager kept $25. The women got back $3 ($1 each). So, far this adds up to $25 + $3 = $28. The remaining $2 was, of course, pocketed by the bellhop. Again, there is no missing dollar.

The following deception puzzle, penned by Lewis Carroll (1958b: 49), plays on the nature of logical deduction itself:

> *I have two clocks: one doesn't run at all and is stuck at 12:00, and the other loses a minute a day. Which would you prefer?*

Most people would immediately pick the clock that loses a minute a day, since, they would argue, at least it runs, whereas the other one does not run at all. But Carroll would disagree. To grasp his logic we start by labeling clocks as follows: (1) A = *a good clock;* (2) B = *the clock that doesn't run at all;* (3) C = *the clock that loses a minute a day.* Compare clock C to clock A. At 12:00 midnight of the first day the two clocks are synchronized. After that first day, the bad clock (C) will be off by one minute at midnight—showing 11:59. After the second day, it will be off by two minutes at midnight—showing 11:58. And so on:

Day	Clock A	Clock C
1	12:00	11:59
2	12:00	11:58
3	12:00	11:57
. . .	. . .	. . .
60th	12:00	11:00

So, after the sixtieth day, the bad clock will be off by one hour at midnight—showing 11:00. Continuing in this way, we can conclude that it will be off by an additional hour after each subsequent sixty-day period:

Day	Clock A	Clock C
61	12:00	10:59
62	12:00	10:58
63	12:00	10:57
...	...	...
120	12:00	10:00

Clearly, the bad clock will have to lose twelve hours in order to become synchronized once again with the good clock. Since it takes it sixty days to lose one hour, it will need 60 × 12 = 720 days to become synchronized again. That is just under two years' time. So it will take almost two years for the clock that loses one minute a day to show the correct time again. The stopped clock B, on the other hand, shows the correct time twice a day—at noon and at midnight. The stopped clock, Carroll concluded, appears logically to be the better one!

Paradoxes

As mentioned at the start of this chapter, the word "paradox" was coined to describe Zeno's puzzling scenarios. In general, a paradox can be defined as any statement or statements that lead to some inconsistency or circularity. The Greek philosopher Protagoras (c. 480–c. 411 B.C.), who was the first thinker to call himself a Sophist, and a teacher of rhetoric named Gorgias (c. 485–c. 380 B.C.) were both clever makers of a type of paradox known as the *self-contradicting liar paradox*. But the most famous of all liar paradoxes is attributed to a Cretan named Epimenides in the sixth century B.C., about whom almost nothing is known:

"All Cretans are liars." Do I speak the truth?

Assume the statement to be true. We can then conclude that Epimenides, being himself a Cretan, was lying. But if he was lying, the statement is not true, and we began by assuming that it was. We have reached a contradiction, an inconsistency. Therefore, our assumption is not correct. The statement must be false: Cretans are not liars. Epimenides is thus a truth-teller. But, then, why did he make a false statement? It is clearly impossible to determine if Epimenides spoke the truth!

This paradox has fascinated logicians and philosophers throughout history. As Eugene Northrop (1944: 12) aptly puts it:

The case of the self-contradicting liar is but one of a whole string of logical paradoxes of considerable importance. Invented by the early Greek philosophers, who used them chiefly to confuse their opponents in debate, they have in more recent times served to bring about revolutionary changes in ideas concerning the nature of mathematics.

An interesting version was invented by British mathematician P. E. B. Jourdain in 1913:

> *On one side of a card is printed the statement "The sentence on the other side of this card is true." But on the card's other side the statement reads "The sentence on the other side of this card is false."*

Jourdain's version is engaging because it materially demonstrates the Janus-faced nature of logical circularity. By making us physically go back and forth between its sides, Jourdain's clever card constitutes a physical model of paradoxical reasoning.

The English philosopher Bertrand Russell found the Liar Paradox to be especially troubling. His restatement of it, known as the Barber Paradox, has animated many debates in modern-day logical circles:

> *The village barber shaves all and only those villagers who do not shave themselves. So, shall he shave himself?*

The barber is "damned if he does and damned if he doesn't," as the colloquial expression goes. If he does not shave himself, he ends up being an unshaved villager. But this goes contrary to the condition set by the puzzle. If he shaves himself, as a villager he must be classified as someone who does not shave himself. But he does shave himself! It is not possible, therefore, for the barber to decide whether or not to shave himself.

Russell argued that such circularity arises because the barber is himself a member of the set of all those who do not shave themselves. So, he concluded, in a consistent logical system the problem of circularity is solved by disallowing such statements about or by a member of the set. Such statements are problematic because they are "self-referential." Russell thus introduced the concept of *metalanguage* in logic—a language "apart from" other language—that was intended to immunize logical systems against the ravages of circularity (Rósza 1957: 258–265). But a serious problem emerged with Russell's solution shortly after he formulated it. It became transparently obvious that metalanguages themselves produce self-referential paradoxes. Russell therefore assumed the existence of

increasingly abstract metalanguages. But each level of metalinguistic reasoning never fails to produce its own self-referential statements. The hier-
archy of metalanguages turns out to be infinite and, hence, totally useless!
Self-referential statements are unavoidable within any type of logical system, with or without a metalanguage.

For some reason, paradoxical statements have a bizarre appeal to all
but diehard Aristotelian logicians. They are especially alluring to children,
as Henry Dudeney (1958a: 15) perceptively observed:

> A child asked, "Can God do everything?" On receiving an affirma-
> tive reply, she at once said: "Then can He make a stone so heavy
> that He can't lift it?"

The child's question is similar to a well-known philosophical conundrum:
*What would happen if an irresistible moving body came into contact with
an immovable body?* As Dudeney observes, such bizarre paradoxes arise
only because we take delight in inventing them. In actual fact, "if there
existed such a thing as an immovable body, there could not at the same
time exist a moving body that nothing could resist."

In their 1986 book *The Liar,* a mathematician named Jon Barwise and
a philosopher named John Etchemendy dismissed such paradoxes because, they asserted, they arise only when statements are not tied to
real-life contexts. So, for instance, when Epimenides says, "All Cretans
are liars," he may be doing so simply to confound his interlocutors. His
statement may also be the result of a slip of the tongue. Whatever the case,
the intent of Epimenides' statement can only be determined by assessing
the context in which it was uttered along with Epimenides' reasons for
saying it. Once such factors are determined, no paradox arises. Actually,
in the 1960s, an attempt had already been made by Professor Lofti Zadeh
of the University of California–Berkeley to incorporate the "pragmatic"
aspect of statements directly into logic. Zadeh claimed that the *fuzzy logic*
that could handle such matters would classify Epimenides' statement as a
"half truth" or a "half falsehood," depending on context, and the barber's
statement as "true under some conditions," but "false under others"
(Stewart 1997: 163). But fuzzy logic has hardly solved the dilemma of the
liar paradox. It has dodged it in a clever manner by bringing "real life"
into the picture. But this raises a series of related paradoxical questions:
What is real life? Why do people lie? And so on and so forth.

A more useful approach would be simply to outlaw formulae, propositions, or procedures within logical systems that lead to inconsistencies
or circularities. Such an approach has, in fact, already been taken by

mathematicians in prohibiting division by 0. Permitting that operation would lead to contradictory results, such as 2 = 1:

Assume that:	$a = b$
Multiply both sides by a:	$a^2 = ab$
Subtract b² from both sides:	$a^2 - b^2 = ab - b^2$
Factor both sides:	$(a + b)(a - b) = b(a - b)$
Divide both sides by (a - b):	$a + b = b$
We started off by assuming a = b, so:	$b + b = b$
Therefore:	$2b = b$
Or:	$2 = 1$

We have thus proven that 2 = 1.

This anomaly arises because we started off by assuming $a = b$, which means that $a - b = 0$. When we divided the equation $(a + b)(a - b) = b(a - b)$ by $(a - b)$, we were dividing by 0. The reason for prohibiting division by 0 is thus a practical one—it is better to retain a system of arithmetic developed over millennia, and proven to be highly useful in everyday life, than to throw it completely out, as the Aristotelian requirement of consistency would entail, because one of the numbers within it is highly problematic. Mathematical life goes on without division by zero. So, too, logical life will go on without self-referential statements.

Paradoxes warn us, as do riddles and optical illusions, that truth is elusive. They also warn us that in generalizing from specific facts we run into difficulties. As Kasner and Newman (1940: 219) aptly put it, "In the transition from *one* to *all*, from the specific to the general, mathematics has made its greatest progress, and suffered its most serious setbacks, of which the logical paradoxes constitute the most important part." As the American philosopher Charles Peirce (1923: 237) also argued, reasoning in generalities is fraught with pitfalls. He demonstrated one such pitfall as follows:

> If a man had to choose between drawing a card from a pack containing 25 red cards and a black one, or from a pack containing 25 black cards and a red one; and if the drawing of a red card were destined to transport him to eternal felicity and that of a black one to consign him to everlasting woe, it would be foolish to deny that he ought to prefer the pack containing the larger portion of red cards, although from the nature of the risk, it could not be repeated. It is not easy to reconcile this with our analysis of the conception of chance. But suppose he would choose

the red pack and should draw the black card. What consolation would he have? He might say that he had acted in accordance with reason, but that would only show that his reason was absolutely worthless. And if he should choose the red card, how could he regard it as anything but a happy accident? He could not say that if he had drawn from the other pack he might have drawn the wrong one, because an hypothetical proposition such as 'If A, then B' means nothing with reference to a single case.

The belief that science and mathematics would be able to explain all mysteries, miracles, and hidden patterns, and therefore that the past and the future could be contained, developed in the post-Galilean era of Western intellectual history. But nothing has proven to be farther from the truth. As products of the puzzle instinct, paradoxes warn us to this day against complacency about our own mental creations. Like a court jester, the *puzzle homunculus* in our brain is always prepared to contrive some paradox that will mischievously undermine our most elaborate logical structures.

As mentioned several times above, paradoxes have had a profound impact on the development of mathematics. The English scientist Sir Isaac Newton (1642–1727) and the German philosopher and mathematician Gottfried Wilhelm Leibniz seem to have been contemplating Zeno's runner paradox (discussed above) when they came up, independently, with an ingenious, yet remarkably simple, solution to it. They simply asserted that the sum to which the series $\{\frac{1}{2} + \frac{1}{4} + \frac{1}{8} + \frac{1}{16} + \ldots\}$ converges as it approaches infinity is the distance between the starting line and the finish line. Thus, the *limit* of the runner's movement is the unit distance of 1. This simple notion became the basis for establishing the calculus.

The idea of limits was not new. Indeed, Archimedes had already used it to calculate the area of a circle. He did this simply by inscribing a regular polygon in a circle. The difference between the perimeter of the polygon and the circumference of a circle, he argued, could be made as small as one desired by progressively increasing the number of sides of the polygon. The limiting figure of such an incremental procedure was the circle, and the limiting area, the area of the circle. One will never be able to calculate the circle's area exactly, Archimedes observed, but one can approximate it as accurately as one wishes.

It is beyond the scope of the present treatment to discuss the historical roots of the calculus. Suffice it to say that it brought about a radical reconsideration of philosophical and religious ideas about the world.

Indeed, when the calculus was first proposed it met with acerbic criticism from philosophers and religious leaders. The Irish prelate and philosopher George Berkeley (1685–1753), for instance, charged that it was a useless science because it dealt with small, meaningless quantities. But the calculus survived such attacks because it produced answers to classical problems of physics. Amazingly, a train of thought initiated by an apparently trivial paradox invented by an ancient Greek philosopher led to the establishment of a science that has allowed human beings to learn a great deal about the universe and, more importantly, about themselves.

The Paradox of Logic

Since antiquity, humanity has prided itself on being a logical species. Legend has it that the Greek philosopher Parmenides invented logic as he sat on a cliff meditating about the world. The English philosopher Thomas Hobbes (1588–1679) claimed that logic was the only attribute that kept human beings from regressing to wild beasts—a view developed further by the French philosopher René Descartes (1596–1650), who refused to accept any belief, even the belief in his own existence, unless he could prove it to be logically true. Descartes also believed that logic was the only way to solve all human problems. In their insightful book, *Descartes' Dream*, Davis and Hersh (1986: 7) encapsulated Descartes's vision as "the dream of a universal method whereby all human problems, whether of science, law, or politics, could be worked out rationally, systematically, by logical computation." Leibniz forged the conceptual link between logic and life even further. Calling logical reasoning a *lingua universalis* (a "universal language" of mind), Leibniz claimed that it could be used to great advantage in the betterment of the human condition for the simple reason that errors in thinking could be reduced to errors in logic and thus easily fixed.

The list of influential thinkers who were inspired by Descartes's dream is impressive indeed. The English logician George Boole drafted a logical system that he claimed would be capable of representing the main laws of thought; the German philosopher Gottlob Frege developed the Boolean system further; and Bertrand Russell, together with Alfred North Whitehead, axiomatized the Boolean-Fregean system into a set of propositions and a metalanguage which, as we saw above, Russell thought would be impervious to the challenges posed by paradoxes. But the Cartesian dream was shattered once and for all by Kurt Gödel in 1931. Motivated by the liar paradox, Gödel showed that logical systems are incomplete, because they invariably contain a statement ("I am not provable") which is *undecidable* in the system, i.e., neither provable nor refutable. This was a truly startling demonstration, and its repercussions are felt throughout mathematics and philosophy to this day.

But even before Gödel, the German philosopher Friedrich Nietzsche (1844–1900) considered the categories of classical logic unreliable tools for gaining objective knowledge about the world because, he warned, they are indistinguishable from the linguistic categories in which they are framed. Moreover, Nietzsche quipped, they are convoluted—a vexatious feature of logic also brought out satirically by Tweedledee in Carroll's *Through the Looking-Glass* with the following remark: "if it was so, it might be; and if it were so, it would be; but as it isn't, it ain't. That's logic."

As Rucker (1987: 218) aptly observes, the "great dream of rationalism has always been to find some ultimate theory that can explain *everything.*" Perhaps what makes paradoxes so mischievously appealing is that they reveal ultimately why a "theory of everything" is impossible. They show that human systems, no matter how commonsensical they may appear to be, are ultimately imperfect. Gödel's wonderful theorem showed, in effect, that logic was made by imperfect logicians, and thus that the Cartesian dream of using logic to solve all human problems was illusory and ultimately empty.

5 Puzzling Numbers

Magic Squares, Cryptarithms, and Other Mathematical Recreations

Mathematics may be defined as the subject in which we
never know what we are talking about, nor whether what
we are saying is true.
                              —Bertrand Russell (1872–1970)

As we saw in the opening chapter, some puzzles have played a
significant role in the development of mathematics. Among the first
works dealing with the basics of mathematical science one finds anthologies of puzzles. The Rhind Papyrus, as we saw, was most likely designed
as an educational tool for teaching problem-solving, as a reference manual in practical mathematical theory, and as a source of brain-puzzling
recreation all in one. And some of the greatest mathematicians of history have investigated theorems and methods through the medium of
puzzles. One of these was Archimedes. Legend has it that he devised his
Cattle Problem to take revenge on one of his adversaries, whom he was
trying to dumbfound with his mathematical prowess. Nevertheless, the
Cattle Problem stimulated the development of notational standards in
mathematics. Cognizant of its potential impact on method, Archimedes
dedicated it to his friend, the great Alexandrian astronomer Eratosthenes.
The original statement of the puzzle is lost. Of the various versions that
have come down to us, the one reproduced below, taken from the
authoritative English-language edition of Archimedes' works by T. L.
Heath (1958: 319), contains the enigmatic extra conditions that follow
the ellipsis:

If thou art diligent and wise, O Stranger, compute the number of cattle of the Sun, who once upon a time grazed on the fields of the Thrinician isle of Sicily, divided into four herds of different colours, one milk white, another glossy black, the third yellow, and the last dappled. In each herd were bulls, mighty in number according to these proportions: understand, stranger, that the white bulls were equal to a half and a third of the black together with the whole of the yellow, while the black were equal to the fourth part of the dappled and a fifth, together with, once more, the whole of the yellow. Observe further that the remaining bulls, the dappled, were equal to a sixth part of the white and a seventh, together with all the yellow. These were the proportions of the cows: the white were precisely equal to the third part and a fourth of the whole herd of the black; while the black were equal to the fourth part once more of the dappled and with it a fifth part, when all, including the bulls, went to pasture together. Now, the dappled in four parts were equal in number to a fifth part and a sixth of the yellow herd. Finally the yellow were in number equal to a sixth part and seventh of the white herd. If thou canst accurately tell, O stranger, the number of cattle of the Sun, giving separately the number of well-fed bulls and again the number of females according to each colour, thou wouldst not be called unskilled or ignorant of numbers, but not yet shalt thou be numbered among the wise . . .

But come, understand also all these conditions regarding the cows of the Sun. When the white bulls mingled their number with the black, they stood firm, equal in depth and breadth, and the plains of Thrinicia, stretching far in all ways, were filled with their multitude. Again, when the yellow and dappled bulls were gathered into one herd they stood in such a manner that their number, beginning from one, grew slowly greater till it completed a triangular figure, there being no bulls of other colours in their midst nor none of them lacking.

If thou art able, O stranger, to find out all these things and gather them together in your mind, giving all the relations, thou shalt depart crowned with glory and knowing that thou hast been adjudged perfect in this species of wisdom.

Today, Archimedes' complicated puzzle is solved in as straightforward a manner as any other word problem in algebra. But it is mind-boggling to think how the ancient Greeks would have gone about solving it, possessing only the most rudimentary of representational techniques. Using modern algebraic notation, we start our solution by letting upper-case X, Y, Z, and T stand for the number of white, black, dappled, and yellow bulls, respectively; and lower-case x, y, z, and t for the corresponding cows. The statements in the puzzle yield seven equations in eight unknowns. These equations are shown below, with fractions added and simplified. Readers can check each equation against the corresponding statement in the puzzle to ascertain its validity:

$$X - T = \tfrac{5}{6} Y$$
$$Y - T = \tfrac{9}{20} Z$$
$$Z - T = \tfrac{13}{42} X$$
$$x = \tfrac{7}{12} (Y + y)$$
$$y = \tfrac{9}{20} (Z + z)$$
$$z = \tfrac{11}{30} (T + t)$$
$$t = \tfrac{13}{42} (X + x)$$

The answer is shown below. Interested readers can consult Dörrie (1965: 4) or Beiler (1966: 249–252) for a comprehensive discussion of this problem:

white bulls	10,366,482
black bulls	7,460,514
dappled bulls	7,358,060
yellow bulls	4,149,387
white cows	7,206,360
black cows	4,893,246
dappled cows	3,515,820
yellow cows	5,439,213

Given the magnitude of the numbers and the complexity of the solution, it is little wonder that in classical antiquity any difficult problem was characterized as a *problema bovinum* ("cattle problem") or a *problema Archimedis* ("Archimedean problem"). Above all else, this puzzle pointed out dramatically how crucial a system of representation is to mathematics. In effect, Archimedes created a puzzle that pinpointed a specific need in the mathematical practices of his times—a need that was filled centuries later with the advent of standardized algebraic notation and method.

The list of puzzles that have led to some mathematical discovery, or have been instrumental in the establishment of a new branch of mathematics, is a long and impressive one indeed. Archimedes' problem sensitized the ancient mathematicians to the need of a useful abstract notation; Fibonacci's Rabbit Puzzle (chapter 1) led to the discovery of many patterns hidden among the integers; Euler's Königsberg's Bridges Puzzle (chapter 1) led to the development of network theory, combinatorics, and topology; and the list could go on and on. As Kasner and Newman (1940: 156) emphasize in their delightful book, *Mathematics and the Imagination,* the "theory of equations, of probability, the infinitesimal calculus, the theory of point sets, of topology, all have grown out of problems first expressed in puzzle form." It is therefore somewhat surprising to find that a purely *recreational* or *enigmatological* branch of mathematics took so long to develop, given the importance of puzzles to the evolution of the discipline. Although individuals such as Alcuin, Fibonacci, Tartaglia, and Cardano wrote mathematical puzzles for various practical and pedagogical reasons in their eras, it was not until 1612 that recreational mathematics emerged as a kind of semi-autonomous branch of mathematics. In that year, the French poet and scholar Claude-Gaspar Bachet de Mézirac published the first comprehensive collection of mathematical puzzles, a book titled *Amusing and Delightful Number Problems* (chapter 1).

Since then, many have come forward to develop and cultivate this branch. Of these, the names of Lewis Carroll, Sam Loyd, François Edouard Anatole Lucas, Henry E. Dudeney, W. W. Rouse Ball, Hubert Phillips, Raymond Smullyan, and Martin Gardner stand out prominently, as we saw in the opening chapter. More recent recreational mathematicians have built on their work and implanted their craft firmly into the broader mathematical landscape (Ahrens 1921; Trigg 1987). In the twentieth century, recreational mathematicians established their own scholarly journals and sources of information, e.g., *Sphinx* (1931–1939), the *Recreational Mathematics Magazine* (1961–1964), the *Journal of Recreational Mathematics* (1968–), and *Eureka* (1978–). And, since the 1950s, they have produced a truly vast repertory of stimulating and challenging puzzles.

On this leg of our journey through Puzzleland, the focus is on the puzzle instinct's fascination with numerical patterns and with the broader implications that such patterns have for an understanding of the world around us. This is the territory inhabited by the Carrollian figure of the Queen of Hearts—the name used, incidentally, to refer to mathematics generally, because mathematics, like a queen, reigns over the whole domain of science.

Magic Squares

One of the oldest and most admired of all puzzle genres is the *magic square*. A magic square is formed by arranging consecutive numbers in the cells of a square array so that the sums of the rows, columns, and diagonals are equal. For example, the numbers 1 to 9 can be arranged into a 3×3 square in this way:

8	1	6
3	5	7
4	9	2

Figure 5.1

Each of the rows, columns, and diagonals of this square adds up to 15 (Andrews 1960: 2). This is called a magic square of *order* 3 (the *order* is the number of rows and columns), and the constant sum—in this case, 15—is called the *magic-square constant*.

There are several other ways to arrange the first nine numbers in nine cells to create a magic square of order 3. This is left as an exercise for the reader (see solution 5.1 for two other possibilities). Obviously, the higher the order of the square, the greater the number of solutions. For example, a magic square of order 5—i.e., a magic square made up of the first 25 numbers—has many more solutions than the one above. The following arrangement, with a magic-square constant of 65, is one such solution (Benson and Jacoby 1976: 4):

17	24	1	8	15
23	5	7	14	16
4	6	13	20	22
10	12	19	21	3
11	18	25	2	9

Figure 5.2

Magic squares were invented in China around 2200 B.C., where they were called *lo-shu*. According to legend, the magic square was a gift from a turtle living in the River Lo to the Emperor Yu the Great, who controlled the flow of the Lo and Yellow Rivers. Given the mystical origins of magic squares, it is little wonder that the ancient Chinese carried with them amulets and talismans with the squares inscribed in them. Magic squares spread from China to the Western world around A.D. 130. A philosopher named Theon of Smyrna in that era was the first Westerner to mention them. Perceiving occult properties in them, ninth-century Arab astrologers used magic squares in casting horoscopes. Four centuries later, around 1300, the Greek mathematician Manuel Moschopoulos spread them throughout the medieval world (Costello 1996: 3). Many considered them to be reifications of hidden numerical patterns that governed the universe. The eminent astrologer Cornelius Agrippa (1486–1535) believed that a magic square of one cell, i.e., a square containing the single digit 1, represented the eternal perfection of God. Agrippa also took the fact that a 2×2 magic square cannot be constructed to be proof of the imperfection of the four elements: air, earth, fire, and water.

As it turns out, the properties of magic squares are truly mysterious and enigmatic. For instance, if the successive terms of the Fibonacci series (chapter 1), starting at 3 and ending with 144—i.e., 3, 5, 8, 13, 21, 34, 55, 89, 144—are matched in order with the integers from 1 to 9 in the above 3×3 square, a new square is formed in which the sum of the products of the three rows equals the sum of the products of the three columns (Pappas 1989: 87):

Original Magic-Square Integer   Replaced with a Fibonacci Number

Original Magic-Square Integer	Replaced with a Fibonacci Number
1	3
2	5
3	8
4	13
5	21
6	34
7	55
8	89
9	144

8	1	6
3	5	7
4	9	2

$\longrightarrow$

89	3	34
8	21	55
13	144	5

Figure 5.3

Row Products		Column Products	
$89 \times 3 \times 34 =$	9078	$89 \times 8 \times 13 =$	9256
$8 \times 21 \times 55 =$	9240	$3 \times 21 \times 144 =$	9072
$13 \times 144 \times 5 =$	9360	$34 \times 55 \times 5 =$	9350
	Sum: 27,678		Sum: 27,678

Patterns such as these, which seem to have no *raison d'être* but which are nonetheless interesting, impart an aura of mystery to the very squares themselves. No wonder, then, that magic squares have caught the fancy of philosophers and artists throughout history. Beguiled by their mysterious charm, the great German Reformation artist Albrecht Dürer (1471–1528) included a picture of a 4×4 magic square, with a magic-square constant of 34, in his famous 1514 engraving *Melancholia:*

16	3	2	13
5	10	11	8
9	6	7	12
4	15	14	1

Figure 5.4

Dürer's writings show that he was captivated by the mysterious patterns that existed among numbers. His painting is, in fact, a metaphor for the supernatural forces that govern the universe. Mesmerized by Dürer's square, Leonhard Euler studied it carefully, showing over two centuries later that there are 880 ways to construct a fourth-order magic square.

Perhaps the most extraordinary of all magic squares was devised by Benjamin Franklin (1706–1790), the great American public official, writer, scientist, and printer:

52	61	4	13	20	29	36	45
14	3	62	51	46	35	30	19
53	60	5	12	21	28	37	44
11	6	59	54	43	38	27	22
55	58	7	10	23	26	39	42
9	8	57	56	41	40	25	24
50	63	2	15	18	31	34	47
16	1	64	49	48	33	32	17

Figure 5.5

Constructed with the first 64 integers, Franklin's 8×8 square contains a host of mind-boggling numerical patterns, which readers can check for themselves, such as the following: (1) its non-diagonal magic-square constant is 260; and exactly half this number, 130, is the constant of each of the four 4×4 squares that are quadrants of the larger square; (2) the sum of any four numbers equidistant from the center is also 130; (3) the sum of the numbers in the four corners plus the sum of the four center numbers is 260; (4) the sum of the four numbers forming any little 2×2 square within the main square is 130. It is truly perplexing to contemplate how Franklin could have ever devised this masterpiece of the genre (Gullberg 1997: 209–212).

From a purely mathematical standpoint, the integers in a magic square are the consecutive numbers from 1 to n^2, where n is the order of the square, i.e., the number of rows and columns. The magic constant—what each line and diagonal adds up to—is representable as follows:

$$\frac{n\,(n^2 + 1)}{2}$$

So, for example, in the 3×3 square with which we began this discussion, $n = 3$, so $n^2 = 3 \times 3 = 9$. Substituting these values into the formula above yields 15.

Although the square is the oldest and most common one, magic figures have been devised with other geometrical forms. The following is a triangular magic figure in which the digits from 1 to 9 are placed in the circles so that each side of the triangle adds up to 20:

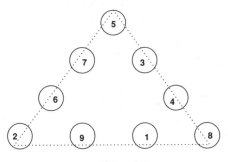

Figure 5.6

Magic pentagrams, hexagrams, circles, and cubes can also be constructed. Magic cubes are particularly interesting because they show that the basic idea underlying magic squares is not limited to two dimensions. They consist of a series of numbers arranged in cubical form so that each row of numbers running parallel with any of the edges, and also any of the four great diagonals, will have the same sum (Benson and Jacoby 1981). The following (partially shown) $3 \times 3 \times 3$ magic cube, with a magic constant of 42, is an example:

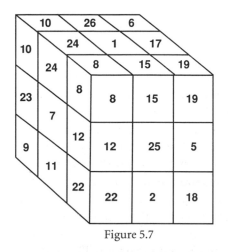

Figure 5.7

Magic figures continue to rouse our curiosity even today because, like the ancients, we continue to be intrigued by pattern in itself. The fascination they hold reflects a deeply embedded feeling that there exists an intrinsic numerico-symbolic connection among things. As Andrews (1960: 162) aptly puts it in his insightful book *Magic Squares and Cubes,* a "study of magic squares may have no practical application, but an acquaintance with them will certainly prove useful, if it were merely to gain an insight into the fabric of regularities of any kind."

Puzzling Mathematical Patterns

Discovering hidden numerical patterns is not the exclusive domain of the creators of magic squares. Every great mathematician seems to have had a knack for doing so. A classic example is provided by a tale told about the German mathematician Carl Friedrich Gauss (1777–1855), who was only ten years old when he purportedly dazzled his teacher with his exceptional computational abilities. One day, his class was asked to cast the sum of all the numbers from one to one hundred: $1 + 2 + 3 + 4 + \ldots + 100 = ?$ Gauss raised his hand within seconds, giving the correct response of 5,050, astounding both his teacher and the other students, who continued to toil over the seemingly gargantuan arithmetical task before them. When his teacher asked Gauss how he was able to come up with the answer so quickly, he is said to have replied (more or less) as follows:

> There are 49 pairs of numbers between one and one hundred that add up to one hundred: $1 + 99 = 100$, $2 + 98 = 100$, $3 + 97 = 100$, and so on. That makes 4,900, of course. The number 50, being in the middle, stands alone, as does 100, being at the end. Adding 50 and 100 to 4,900 gives 5,050.

Now, while Gauss's method may have dazzled his teacher and his peers, it comes as little surprise to anyone familiar with the notion of *sequence* or *series*. What Gauss carried out, in effect, was a summation of the sequence $1 + 2 + 3 + \ldots + 100$. Using the language of algebra, the problem can be expressed more generally as, What is the sum of $1 + 2 + 3 + \ldots + n$, where n is any whole number? The answer is $n(n + 1)/2$. Substituting 100 for n produces the answer 5050.

How did mathematicians arrive at such a formula? They arrived at it by reasoning exactly as Gauss did. The sum of the first n numbers can be written as follows: $Sum = 1 + 2 + 3 + \ldots + n$. The same sum can be written with the digits in reverse order. This changes nothing in the result: $Sum = n + (n - 1) + (n - 2) + \ldots + 2 + 1$. We can now line up these two equivalent versions of the same sum, as shown below:

(1) $Sum = 1 + 2 + \ldots + (n - 1) + n$
(2) $Sum = n + (n - 1) + \ldots + 2 + 1$

By adding (1) and (2) we will get, on the right-hand side, the expression $(n + 1)$ each time we add up a pair of corresponding expressions:

$$1 + n = (n + 1)$$
$$2 + (n - 1) = (n + 1)$$
$$3 + (n - 2) = (n + 1)$$

and so on.

Such additions occur exactly n times. Therefore, the result of adding up (1) and (2) is: $2Sum = n(n + 1)$. This can be simplified to $Sum = n(n + 1)/2$.

Numerical pattern has been the target not only of the great mathematical theorists but also of many resourceful puzzlists, from Ahmes to present-day recreational mathematicians. The following is an example of a multiplication pattern disguised as a military problem, from the pen of an Indian scholar of the mid-ninth century named Mahavira, who wrote quite a number of interesting mathematical puzzles (Mahavira 1912: 112):

> *Three* puranas *formed the pay of one man who is a mounted soldier; and at that rate there were sixty-five men in all. Some (among them) broke down, and the amount of their pay was given to those that remained in the field. Of this, each man obtained 10* puranas. *You tell me, after thinking well, how many remained in the field and how many broke down.*

Mahavira's puzzle asks us, essentially, to complete a pattern that can be expressed in arithmetical terms as follows: *What two numbers, both less than 65, would produce the same product when one is multiplied by 3 and the other by 10?* In the words of the puzzle, the first number represents the soldiers who broke down, each of whom was paid 3 *puranas,* and the second the soldiers who remained in the field, each of whom received 10 *puranas.* The two numbers are 15 and 50, since 3 *puranas* × 50 = 150 *puranas* and 10 *puranas* × 15 = 150 *puranas.* Thus, 50 soldiers broke down and 15 remained in the field.

The next puzzle, from Alcuin's *Problems to Sharpen the Young* (chapter 1), deals with a different kind of pattern, now investigated under the rubric of *critical path analysis* (Holt and McIntosh 1966: 9). Dubbed the River-Crossing Puzzle, it has become a fixture of recreational mathematical collections since Alcuin (see Ascher 1990 on the appearance of this puzzle in other parts of the ancient world):

> *A traveler comes to a riverbank with a wolf, a goat, and a head of cabbage. There is only one boat for crossing, which can carry no more than the traveler and one of the two animals or the cabbage. Unfortunately, if left alone together, the goat will eat the cabbage, and the wolf will eat the goat. How does the traveler transport his animals and his cabbage to the other side safely and soundly in a minimum number of trips?*

The initial situation (the 0 position) on both sides of the river, before the traveler starts ferrying items back and forth, can be shown as follows (W = wolf, G = goat, C = cabbage, T = traveler):

Original Side	On the Boat	Other Side
0. W G C T	— —	— — —

Since he can only take himself and one thing (cabbage or animal) with him, the traveler can start by transporting the goat over on the boat. This leaves the wolf and the cabbage alone on the original side without any problems, because the wolf does not eat cabbage:

Original Side	On the Boat	Other Side
0. W G C T	— —	— — —
1. W — C	T G →	— — —

The traveler deposits the goat on the other side and then goes back alone. This completes the first round trip:

Original Side	On the Boat	Other Side
0. W G C T	— —	— — —
1. W — C	T G →	— — —
2. W — C	← T	— G —

From the original side, the traveler can now take the wolf across with him:

Original Side	On the Boat	Other Side
0. W G C T	— —	— — —
1. W — C	T G →	— — —
2. W — C	← T	— G —
3. — — C	T W →	— G —

The traveler cannot leave the wolf and goat alone while he goes back for the cabbage, for the former would eat the latter. So he must bring the goat back for the ride, leaving the wolf harmlessly by itself:

Original Side	On the Boat	Other Side
0. W G C T	— —	— — —
1. W — C	T G →	— — —
2. W — C	← T	— G —
3. — — C	T W →	— G —
4. — — C	← T G	W — —

Back on the original side, the traveler can then leave the goat and take the cabbage with him over to the wolf:

Original Side	On the Boat	Other Side
0. W G C T	— —	— — —
1. W — C	T G →	— — —
2. W — C	← T	— G —
3. — — C	T W →	— G —
4. — — C	← T G	W — —
5. — G —	T C →	W — —

He can thereupon go back alone, leaving the wolf and the cabbage on the other side with no disastrous consequences:

Original Side	On the Boat	Other Side
0. W G C T	— —	— — — —
1. W — C	T G →	— — — —
2. W — C	← T	— G —
3. — — C	T W →	— G —
4. — — C	← T G	W — — —
5. — G —	T C →	W — — —
6. — G —	← T	W — C

For his last trip (his seventh), the traveler brings the goat over and then continues happily on his journey with his wolf, goat, and cabbage:

Original Side	On the Boat	Other Side
0. W G C T	— —	— — — —
1. W — C	T G →	— — — —
2. W — C	← T	— G —
3. — — C	T W →	— G —
4. — — C	← T G	W — — —
5. — G —	T C →	W — — —
6. — G —	← T	W — C
7. — — — —	T G →	W — C

A more elaborate version of Alcuin's puzzle was put forward by the eminent Renaissance mathematician Niccoló Tartaglia (Kasner and Newman 1940: 159):

> *Three beautiful brides with their jealous husbands come to a river. The small boat which is to take them across holds only two people. To avoid any compromising situations, the crossings are to be so arranged that no woman shall be left with a man unless her husband is present.*

In this case nine trips are required, with different wives going back and forth (see solution 5.2). Complex versions of the River-Crossing Puzzle have been instrumental in the development of critical path analysis and, more generally, of combinatorics (the study of combinatory pattern), which has, in turn, become important in probability theory and statistics, in the design and operation of computers, and in the physical and social sciences.

The arithmetical puzzle below, from the pen of Tartaglia, is concerned not with unraveling pattern, but with mischievously hiding it:

A man dies, leaving 17 camels to be divided among his heirs in the proportions ½, ⅓, ⅑. How can this be done?

Dividing up the camels in the manner decreed by the father would entail having to split up a camel. This would, of course, kill it. So Tartaglia suggested "borrowing an extra camel," for the sake of mathematical argument, not to mention humane purposes. With 18 camels, we arrive at a practical solution: one heir was given ½ (of 18), or 9, another ⅓ (of 18), or 6, and the last one ⅑ (of 18), or 2. The 9 + 6 + 2 camels given out in this way add up to the original seventeen! The extra camel could then be returned to its owner. Clearly, the clever Tartaglia devised his puzzle as a play on numerical pattern—a pattern obfuscated by the "real-life" conditions presented in the puzzle.

The French puzzlist Bachet devised the following puzzle, based on an enigmatological concept developed by Fibonacci and Tartaglia before him. He too designed it to conceal a simple numerical pattern:

What is the least number of weights that can be used on a scale pan to weigh any integral number of pounds from 1 to 40 inclusive, if the weights can be placed in either of the scale pans?

It may seem at first that six weights are required, of 1, 2, 4, 8, 16, and 32 pounds. Using them, we can weigh 1 pound of, say, rice by putting the 1-pound weight in the left scale pan and pouring rice into the right pan until the pans balance. We weigh two pounds of rice with the 2-pound weight; to weigh 3 pounds, we put both the 1-pound and the 2-pound weight into the left pan; and so on.

However, since the puzzle allows the weights to be put on both sides of the scale, the clever Bachet could get the weighings done with only four weights, of 1, 3, 9, and 27 pounds. The reason for this is remarkably simple; placing a weight in the right pan, with the rice, is equivalent to taking it away from the weight in the left pan. The choice of these four weights rests on the fact that each of the numbers from 1 to 40 is either a multiple of 3 or is one more or one less than a multiple of 3; the weights incidentally are powers of three ($1 = 3^0$, $3 = 3^1$, $9 = 3^2$, $27 = 3^3$). Without going into the details of Bachet's solution here, suffice it to say that his weights will do the job every time because they were selected on the basis of this hidden number pattern (see solution 5.3).

Today Bachet's puzzle would be classified as a problem in *opera-*

tional analysis. In miniature form, it shows the importance of breaking down procedures and studying their parts in order to unravel a principle or pattern that may be hidden within them.

The weighing scale has, as it happens, been the source of innumerable ingenious puzzles, which constitute what may be called simply a *culprit* genre. The following is a typical example of a culprit puzzle:

> *I have six billiard balls, one of which weighs less than the other five. They all look the same. How can I identify the one that weighs less on a balance scale with only two weighings?*

Making a "trial run" with a simpler scenario allows us to see if there is some general principle or procedure that can be applied to solving this puzzle. Consider the weighing of two balls—the simplest scenario of all. Clearly, we can put each ball on a scale pan at the same time—one on the left pan and one on the right. The pan that goes up, of course, is the one holding the ball that weighs less. In this case, one weighing was sufficient to identify the "culprit" ball. Next, consider the weighing of four balls. First, we divide the four balls in half: i.e., into two sets of two balls each. We put two balls on the left pan and two on the right pan. The pan that goes up contains the ball that weighs less, but we do not yet know which one of the two it is. So we take the two suspect balls from the pan that went up, discarding the ones on the other pan. Then we put each one on a separate pan—one on the left pan and one on the right. The pan that goes up identifies the ball that weighs less.

These two trial runs have shown us how to go about methodically identifying the culprit ball in a collection of any even number of balls. We are now ready to turn our attention to the original puzzle. There are six balls in the collection, and we are told to identify the culprit ball in only two weighings. We start off in the same way, dividing the six balls equally in half: i.e., into two sets of three balls each. We put three balls on each pan this time—three on the left pan and three on the right. The pan that goes up contains the ball that weighs less, but we do not yet know which one of the three it is. Now, for the second weighing, we focus our attention on the set containing the culprit ball, discarding the ones on the other pan. We select any two of the three balls to weigh, putting the third ball aside for the time being. We then put each of the two balls on a separate pan—one on the left pan and one on the right. Now, if the pans balance, the culprit ball is the one that we had put aside; if they do not, then the culprit ball is on the pan that goes up. In either case, it takes just two weighings to identify the ball that weighs less.

The same type of solution applies to an odd number of balls (see solution 5.4 for a discussion of a seven-ball version). Not only weighing scales, but jars, cans, and all kinds of containers have been used by puzzlists as practical devices for experimenting with arithmetical patterns and optimal operational procedures of all kinds. One such puzzle, which has become a standard entry in modern-day puzzle anthologies, was penned in the fifteenth century by a certain Nicolas Chuquet, a doctor by profession, who published a remarkable little puzzle book titled *Triad on the Science of Numbers* in 1484 (Wells 1992: 30–31):

> *You have two jars that will hold 5 and 3 pints respectively, neither jar being marked in any way. How can you measure exactly 4 pints from a cask with an unspecified quantity of liquid in it, given that you are allowed to pour liquid back into the cask?*

The solution is left as an exercise for the reader. The steps that the clever Chuquet suggests are given in solution 5.5.

The following truly ingenious liquid measuring puzzle was devised by Lewis Carroll in 1885 (Hudson 1954: 756). It continues to baffle and intrigue solvers to this day and thus merits a detailed discussion here:

> *There are two containers on a table, A and B. A is half full of wine, while B, which is twice A's size, is one-quarter full of wine. Both containers are filled with water and the contents are poured into a third container, C. What portion of container C's mixture is wine?*

Since container A is half full of wine and container B, which is twice the size of A, is one quarter full of wine, the containers have the same quantity of wine in them, as can be seen in the following visual representation of the puzzle [figure 5.8].

Filling the containers with water produces the following proportions [figure 5.9]:

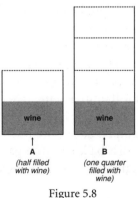

Figure 5.8

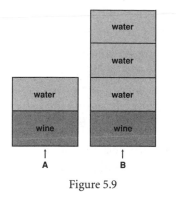

Figure 5.9

A has equal amounts of wine and water, while B has three parts water and one part wine. Between the two containers, there are six equal parts in total—two of wine and four of water. That is, in fact, what container C will have:

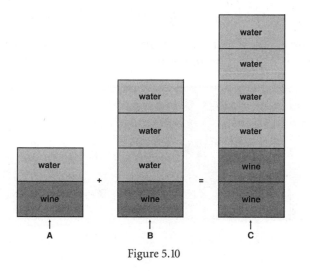

Figure 5.10

The wine and water in container C will, of course, be mixed up, not separated neatly as we have shown them for the sake of convenience in the diagrams above. But in that mixture, wine will make up 2 parts out of 6, or ⅔, and water will make up 4 parts out of 6, or ⅘. In conclusion, C's mixture will contain ⅔ = ⅓ wine.

Many puzzles have also been created as warnings to solvers not to take for granted the given conditions in a situation or in a set of procedures. The following puzzle is an example of this genre. It requires solvers to think of what actually happens when cigarettes are smoked under the stated conditions (see solution 5.6):

> *Mack decided to quit smoking as soon as he finished the 27 cigarettes left in his pocket. Since he habitually smoked only two-thirds of a cigarette at a time, he realized that he could re-roll his butts into new cigarettes. If he smoked only once each day, how many days went by before he finally quit his bad habit?*

The next puzzle plays, instead, on a hidden visual pattern related to orientation. Since it is often found in puzzle collections, and according to the modern-day puzzlist Townsend (1986) is one of the hundred or so most loved puzzles of all time, it merits a detailed discussion here.

> *Three books, each of the same width, are stacked upright against each other on a bookshelf. Each cover is ½ inch thick, and the pages of each book are 2 inches thick. A bookworm starts on the first page of the book on the left and bores its way straight through to the last page of the book on the right. How far has the bookworm gone?*

The three volumes in order from left to right, together with the thicknesses of their covers and pages, can be shown as follows:

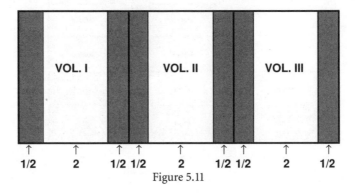

Figure 5.11

Next, we indicate the first page of the first volume and the last page of the last volume [see figure 5.12]. Readers having difficulty seeing this can simply take three books, stack them up as shown, and then see for themselves where the first page of the book on the left is and where the last page of the book on the right is.

We are told that the bookworm started on the first page of the book on the left (where the first arrow is) and ended up on the last page of the book on the right (where the second arrow is) [see figure 5.13].

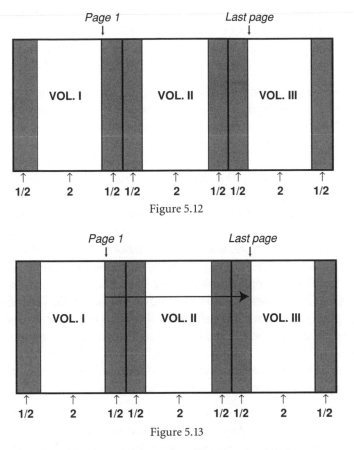

Figure 5.12

Figure 5.13

Now we can easily calculate the distance the bookworm covered: ½ + ½ + 2 + ½ + ½ = 4 inches.

The next puzzle is another classic puzzle that requires a very careful reading of the text:

> *In a box there are 20 balls, 10 white and 10 black. With a blindfold on, what is the least number of balls one must draw out in order to be sure of having a pair of matching balls?*

Some newcomers to this puzzle tend to reason somewhat along the following lines:

> If the first ball that I pull out is white, then I will need another white one to match it. But the next ball might be black, as might be the one after that, and the one after that, and so on. So, in order to be sure that I get a match, I may have to remove all the black balls from

the box—10 in all. The next one I remove after that will then necessarily be white. Including the first white ball I took out, the ten black balls, and the one white ball that matches, 12 is the minimum number of balls I will need to draw out to be sure of getting a match.

This line of reasoning, however, fails to take into account what the puzzle requires us to do—namely, to match the color of any two balls, not just the color of the first one drawn out. Suppose the first ball drawn out is white. If we are lucky, the next ball drawn will also be white, and it's game over. But we cannot assume this best-case scenario. We must, on the contrary, assume the worst-case scenario, namely that the next ball drawn is black. Thus, after two draws, we will have one white and one black ball. Obviously, we could have drawn out a black ball first and a white one second. The end result would have been the same: one white and one black ball. Now, the next ball drawn from the box will, of course, be either white or black. Suppose we draw out a white ball; then it matches the previously drawn white ball. If instead the next ball is black, then it matches the previously drawn black ball. Therefore, no matter what color the third ball is, it will match the color of one of the two already drawn. So the least number of balls we will need to draw from the box in order to ensure a pair of matching balls is three.

Note, however, that if gloves are involved, rather than balls, then the reasoning that was erroneous above is now appropriate:

> If there are six pairs of black gloves and six pairs of white gloves in a drawer, all mixed up, what is the least number of draws that are required in order to guarantee a matching pair of black or white gloves?

Because some gloves fit on the right hand and some on the left hand, one might pick all 12 left-hand gloves, as a worst-case scenario. However, the thirteenth glove will be a right-handed one and also match one of the previous twelve in color. Thus, in this case 13 gloves may have to be drawn in order to get a pair of matching gloves.

Such puzzles are contemporary derivatives of the following ingenious puzzle devised by Lewis Carroll (1958c: 7):

> A bag contains one counter, known to be either white or black. A white counter is put in, the bag shaken, and a counter drawn out, which proves to be white. What is now the chance of drawing a white counter?

We let B and W-1 stand respectively for the black and white counters that might be in the bag at the start, and W-2 for the white counter added to the bag. Removing a white counter from the bag entails three equally likely combinations of two counters, one inside and one outside the bag:

	Inside the Bag	Outside the Bag
(1)	W-1	W-2
(2)	W-2	W-1
(3)	B	W-2

In combination (1), the white counter drawn out is the one that was put into the bag (W-2), and the white counter inside it (W-1) is the counter originally in it. Combination (2) is the converse of (1): the white counter drawn out is the one that was originally in the bag (W-1), and the white counter inside it (W-2) is the counter that was put in. In combination (3), the white counter drawn out is the one that was put into the bag (W-2), since there was no white counter originally in it. The counter that remains in the bag is a black one (B). In two of the three cases, Carroll observes, a white counter remains in the bag. So the chance of drawing a white counter on the second draw is two out of three.

The following puzzle is a version of the same basic idea. Its solution is left as an exercise for the reader (see solution 5.7):

> *Imagine three boxes: one with two black ties in it, another with two white ties in it, and a third with one white and one black tie. The boxes are labeled, logically enough, BB (= two black ties), WW (= two white ties), BW (= one black, one white tie). However, someone has switched the labels, so that now each box is labeled incorrectly. Can you determine the actual contents of each box by drawing out just one tie from one of the boxes?*

The following puzzle, attributed to Martin Gardner (Costello 1996: 119), is yet another version of Carroll's ingenious concept:

> *There are three closed boxes on a table, which contain, separately, 10¢, 15¢, and 20¢ in nickels. However, they are labeled incorrectly. Someone takes the contents out of the box labeled 15¢, two nickels, and puts the nickels out in front of the box. Can you tell the contents of each box?*

This puzzle requires that we match up box labels, say A, B, and C, and their actual contents. First, we show in diagram form what we know: (1) that the boxes contain 10¢ (= two nickels), 15¢ (= three nickels), and 20¢ (= four nickels); (2) that each box is mislabeled, e.g., if it says 10¢,

then we know for certain that it does not have 10¢ in it, but 15¢ or 20¢;
(3) that the contents of box B, labeled 15¢, are two nickels (10¢):

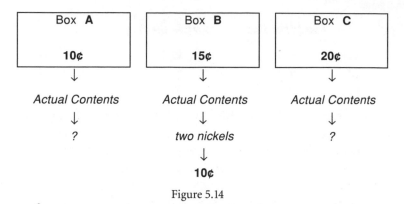

Figure 5.14

We know that one of the remaining two boxes, A and C, contains
three nickels (15¢) and the other, four nickels (20¢). This implies two
possible scenarios [see figure 5.15].

Scenario 1 contradicts a given fact—C is labeled correctly as containing 20¢, contrary to the fact that all three boxes are labeled incorrectly. So we can reject this scenario. Scenario 2, on the other hand, produces no contradictions. Thus, A contains 20¢, B 10¢, and C 15¢.

Algebraic Puzzles

The story of algebra begins with the ancient Babylonians, who introduced the notion of equations and the means of solving them by essentially the same procedures taught today, but without a notational system
for doing so. Ever since, algebra has provided much enigmatological
fodder for puzzlists. The oldest algebraic puzzle known is problem 29 in
the Rhind Papyrus. It can be paraphrased as follows:

> I think of a number, and add to it two-thirds of the number. I then
> subtract one-third of the sum. My answer is 10. What number did
> I think of?

Using modern-day algebraic notation, we can represent the unknown
number with the letter n. Next, we convert each statement into an algebraic expression [see figure 5.16].

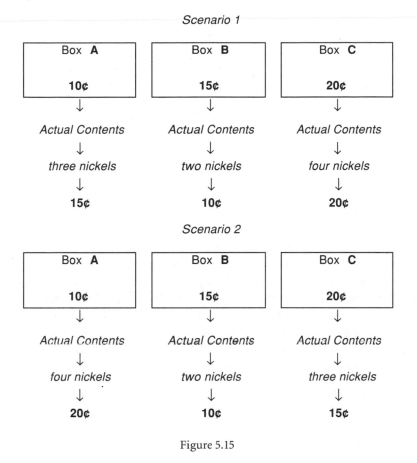

Figure 5.15

Verbal Statement	→	**Algebraic Equivalent**		
Replace "two thirds of the number" with its algebraic equivalent	→	$2/3\,n$		
Add this to the number n	→	$2/3\,n + n$	→	$5/3\,n$
Replace "one third of the above sum" with its algebraic equivalent	→	$1/3\ (5/3\,n)$	→	$5/9\,n$
Subtract this from the sum 5/3 n	→	$5/3\,n - 5/9\,n$	→	$10/9\,n$
Replace "my answer is 10" with its algebraic equivalent	→	$10/9\,n = 10$		

Figure 5.16

Solving the equation $10/9n = 10$, we get $n = 9$. As this solution shows, an algebra puzzle is, in effect, a problem in translation, in which linguistic statements are to be converted into algebraic statements.

The ancient mathematician Diophantus became well known for his algebraic puzzles (chapter 1). Here is an example of his art:

> *What number must be added to 100 and to 20 (the same number to each) so that the sums are in the ratio 3:1?*

We let the number to be added be n; then $(100 + n)$ and $(20 + n)$ represent the numbers that result when it is added respectively to 100 and to 20. Now, we are told that $(100 + n)$ is to $(20 + n)$ as 3 is to 1. This translates into the following equation: $(100 + n)/(20 + n) = 3/1$. Solving the equation, we get $n = 20$.

Diophantus's most challenging puzzles are those in which there are fewer equations than variables. As mentioned, these are known appropriately as Diophantine puzzles (chapter 1). They are solved by keeping in mind the conditions stated in the puzzle that restrict the range of values that can be assigned to the variables (a good collection of such puzzles can be found in Singmaster 1999). Archimedes' Cattle Problem and Alcuin's Grain Puzzle (discussed in chapter 1) are examples of this genre. Here is another Diophantine puzzle (see solution 5.8):

> *Zerlina is a wonderful grandmother. Everybody loves her. Her age is between 50 and 70. Each of her sons has as many sons as brothers. The combined number of Zerlina's sons and grandsons equals her age. How old is Zerlina?*

As such puzzles reveal, algebra is essentially the "science of restoration and balancing," as the medieval Islamic mathematicians characterized it. The Arabic word for "restoration," *al-jabr*, is in fact the root of the word "algebra." In the ninth century, Mohammed ibn-Musa al-Khwarizmi wrote the first true algebra textbook, a systematic account of the basic methods for "restoring" and "balancing" equations. The more difficult algebraic puzzles require, in fact, a solid knowledge of these methods, as the following puzzle emphasizes. The solution is left as an exercise for the reader (see solution 5.9):

> *Two animal-loving friends, Betty and Bill, decided to go on an animal-watching outing last week way out in the country. From a high hilltop, they spotted a flock of sheep down below. Bill cried out, "Look, Betty, there must be a hundred sheep in that flock."*

Betty, who was an accountant and very fastidious about being nu-
merically accurate, but who also loved to stump her friends with
her mathematical wizardry, answered rather matter-of-factly as
follows:
"Nonsense. In order to add up to one hundred, there would
have to be as many again as there are in the flock, and then half as
many more than there are, and then a quarter as many as there
are. And, even then, there would still have to be one more sheep,
before there would be one hundred."
Can you figure out how many sheep there were in the flock?

As we have seen in other domains of Puzzleland, puzzlists have not always had in mind the noble objective of putting on display the power of a method or a notational system. Here is an example of a craftily contrived algebra puzzle that is found in many anthologies of recreational mathematics. Its aim is to dupe the solver into a wrong solution:

A bottle and a cork together cost 55 cents. The bottle costs 50 cents
more than the cork. How much does the cork cost?

Given the way the statement is laid out, many solvers are fooled into thinking that the bottle, costing 50 cents more, must cost 55 cents. But then this means that the cork costs 5 cents and thus that the total cost is 60 cents, contrary to the given facts. The appropriate algebraic statement is $x + x + 50 = 55$ (or $2x = 5$), where x is the price of the cork and $x + 50$ the price of the bottle. Solving the equation, we find that the cork costs 2½ cents.

Cryptarithms

Of all the mathematical puzzle genres discussed in this chapter, the *cryptarithm* is the only one that does not have ancient or medieval origins. As Maxey Brooke (1963a: 3) asserts, the term "cryptarithm" was originally coined by a pseudonymous "M. Vatriquant" in the May 1931 issue of the Belgian magazine *Sphinx*. This genre was not devised, like magic squares, for some mystical reason, but for the sole purpose of providing mental challenge and entertainment. Its creator is generally thought to have been Sam Loyd.

A cryptarithm is a puzzle in which some or all of the digits in an addition, subtraction, division, or multiplication problem have been deleted. The solver is required to reconstruct the problem by deducing

numerical values on the basis of the mathematical relationships indicated by the various arrangements and locations of the given numbers.
Cryptarithms are, in effect, the arithmetical counterparts of cryptograms
(chapter 2). Here is the puzzle with which Loyd (1959–60: 39) introduced
this genre to his many fans.

```
                    * 5 3
            * * 9 | 6 * 8 * * *
                    * * * 2
                    ─────────
                      * 9 * *
                      * * 4 *
                    ─────────
                        * * 4 *
                        * * * *
                      ─────────
```

Figure 5.17

A few numbers can be reconstructed quickly by simple inspection:
(1) the 3 of the quotient, when multiplied by the 9 of the divisor, yields
the product 27, so a 7 can be put in place of the asterisk under the 3; (2)
this means that the last asterisk in the second row from the bottom is
also 7; (3) since there is no remainder, a 4 can be inserted under the 4 in
the second-to-last row, and a 7 under the 7 in the second-to-last row; (4)
the 5 of the quotient, when multiplied by the 9 of the divisor, yields the
product 45, so the 5 digit can be put in place of the last asterisk of the
third-to-last row from the bottom; (5) since a number is subtracted from
8 to leave 9, the only possibility is that a 1 was carried to the 8 to make
18, and the asterisk below it is therefore 9:

```
                    * 5 3
            * * 9 | 6 * 8 * * 7
                    * * 9 2
                    ─────────
                      * 9 * *
                      * * 4 5
                    ─────────
                        * * 4 7
                        * * 4 7
                      ─────────
```

Figure 5.18

Now, the digit above the 5 in the third-to-last row can only be 9, because 5 is subtracted from it and leaves 4. This means that the second

asterisk from the right in the dividend must also be replaced by a 9. Now, since the 3 in the quotient, when multiplied by the 9 in the divisor, yields 27, the 4 in the second-to-last row tells us that the number which 3 will multiply next in the divisor—the middle asterisk—must produce a 2. The only way this can be made to happen is by replacing the divisor's middle asterisk with a 4 ($3 \times 4 = 12$):

$$
\begin{array}{r}
* 5\,3 \\
* 4\,9\ \overline{\smash{)}\ 6 * 8 * 9\,7} \\
* * 9\,2 \\
\hline
* 9 * 9 \\
* * 4\,5 \\
\hline
* * 4\,7 \\
* * 4\,7 \\
\hline
\end{array}
$$

Figure 5.19

The first asterisk in the quotient must be 8, since only 8×9 produces a number ending in 2 (72), which is the final digit in the row below the dividend. This completes the quotient:

$$
\begin{array}{r}
8\,5\,3 \\
* 4\,9\ \overline{\smash{)}\ 6 * 8 * 9\,7} \\
* * 9\,2 \\
\hline
* 9 * 9 \\
* * 4\,5 \\
\hline
* * 4\,7 \\
* * 4\,7 \\
\hline
\end{array}
$$

Figure 5.20

The last step is to complete the divisor by finding its first digit. In long division, our first step would normally be to divide the divisor (*49) into the first three digits of the dividend (6*8). But in the figure, the first step was to divide the divisor into the first *four* digits of the dividend (6*8*). This means that the divisor must be greater than the first three digits of the dividend. The divisor's first digit must therefore be 6, 7, 8, or 9. By trying them in turn, we discover that it must be 7 (749 × 853 = 638,897). At this point the rest of the problem can be reconstructed mechanically.

```
                853
          749 | 638897
                5992
               ‾‾‾‾‾‾
                3969
                3745
               ‾‾‾‾‾‾
                2247
                2247
               ‾‾‾‾‾‾
```

Figure 5.21

In 1924 Henry Dudeney came up with a different genre of crypt-arithm puzzle, in which he replaced all the numbers in an addition, sub-traction, multiplication, or division problem with letters. Subsequently, in 1955, the Canadian puzzlist J. A. H. Hunter referred to this genre of cryptarithm as an *alphametic* (short, obviously, for *alphabet arithmetic*). Here is Dudeney's original addition alphametic:

SEND + MORE = MONEY

The reasoning process involved in solving an alphametic is identi-cal to that used in solving a cryptarithm. We start by establishing that the M at the extreme left of the sum is a carry-over digit equal to 1, be-cause 1 is the only carry-over possible when two digits are added to-gether in the previous column—in this case S + M—even if the column has itself a carry-over from the column before. If readers do not see this, they should consider the column in question. We know that the two digits being added are different, since one is represented as S and the other as M. The most two different digits can add up to is 17; this would occur with the two largest digits, 9 and 8. So let the two digits in the col-umn equal 9 and 8, just for the sake of illustration, ignoring for the moment the actual letters that are there:

	↓			
	9	E	N	D
+	8	O	R	E
1	7	N	E	Y

Now it is obvious that M can only equal 1. Even if there were a carry-over from the previous column, the most 9 + 8 + 1 (carry-over) could possibly add up to is 18. Having proved that M = 1, we put its numerical value in the puzzle, noting that it occurs in two places:

	S	E	N	D
+	1	O	R	E
1	O	N	E	Y

Now, we focus on the column S + 1 = O. We have established that M = 1, so we know that O does not equal 1. It must therefore equal 0 (zero), since the S above it can be no more than 9. Adding together 9 + 1, we get 10. We know that the sum is not 11 because that would make the O below equal to 1, which, as we have just determined, is the value of M, not of O. So we can go ahead and replace O with 0 (zero) in the places where it is found in the layout:

	S	E	N	D
+	1	0	R	E
1	0	N	E	Y

Now consider the S in the column. It must be either 9 or 8. A choice of 8 would mean that there is a carry-over from the previous column. But, looking at that column, we notice that it has a 0 in it: E + 0 = N. So there is no way that this column can produce a carry-over, even if E = 9, since the digit 1 has been already assigned to M. So there can be no carry-over from the E + 0 = N column to the S + 1 = 0 column. We can thus safely conclude that S = 9:

	9	E	N	D
+	1	0	R	E
1	0	N	E	Y

Now we turn our attention to the E + 0 = N column. It must have a carry-over from the previous N + R = E column, because if there were no carry-over, the E + 0 = N column would not make arithmetical sense, since the sum of any number and 0 is the original number. So, without a carry-over, E + 0 should add up to E. But it does not. So E + 0 = N must have a carry-over. We can show this arithmetically as 1 (carry-over) + E + 0 = N.

		1	←	*carry-over*	
	9	E	N	D	
+	1	0	R	E	
1	0	N	E	Y	

The equation 1 + E + 0 = N simplifies to N = E + 1. We store this fact momentarily in memory. Now, consider the N + R = E column. As we have just determined, the addition of N and R produces a carry-over. In arithmetical terms, this means that the sum of N and R, which produces the digit E, is greater than 10. This fact can be represented, of course, with E + 10. So now we know that N + R = E + 10. This assumes, however, that there is no carry-over from the previous column. If there is a carry-over from that column, then the appropriate equation is 1 (carry-over) + N + R = E + 10, which can be simplified to N + R = E + 9.

Recall that we have established that N = E + 1. So we can now substitute this value into the two possibilities for the column:

(1) *With no carry-over:*
N + R = E + 10 → E + 1 + R = E + 10 → R = 9
(2) *With a carry-over:*
N + R = E + 9 → E + 1 + R = E + 9 → R = 8

Therefore R = 9 or 8. We have already established that S = 9. So we conclude that R = 8:

	9	E	N	D
+	1	0	8	E
1	0	N	E	Y

Consider the N + 8 = E column. The digits 0 and 1 have already been assigned, so we can deduce that this column produces a carry-over, since N is at least 2. Now, it is obvious that the D + E = Y column will also produce a carry-over, since N and E cannot be 0 or 1. N also cannot be 2, for then E would be equal to 0 or 1 in the N + 8 = E column, which is impossible since these two digits have already been assigned. Let's try out a few possibilities, starting with N = 3. Hypothetical replacements are shown in italics:

	9	E	*3*	D
+	1	0	8	E
1	0	*3*	E	Y

From the E + 0 = 3 column, we can see that E = 2, since the column has a carry-over. We put this value in the appropriate locations:

		1	←	*carry-over*	
	9	*2*	*3*	D	
+	1	0	8	*2*	
1	0	*3*	*2*	Y	

The $3 + 8 = 2$ column has a carry-over from the ones column. So the D in the first column must be 8 or 9. But this is impossible, since these digits have been already assigned. So we discard our original hypothesis, namely that $N = 3$. We can now try $N = 4$ instead:

$$
\begin{array}{ccccc}
 & 9 & E & 4 & D \\
+ & 1 & 0 & 8 & E \\
\hline
1 & 0 & 4 & E & Y \\
\end{array}
$$

This replacement makes $E = 3$:

$$
\begin{array}{ccccc}
 & & 1 & & \leftarrow \quad \textit{carry-over} \\
 & 9 & 3 & 4 & D \\
+ & 1 & 0 & 8 & 3 \\
\hline
1 & 0 & 4 & 3 & Y \\
\end{array}
$$

Now, D must be equal to 7, 8, or 9. We have already assigned 8 and 9, so these two possibilities can be discarded. We can also discard 7, because if $D = 7$, then $Y = 0$, but 0 has also already been assigned. Once again, we discard our working hypothesis, and go on to $N = 5$ as our next hypothesis:

$$
\begin{array}{ccccc}
 & 9 & E & 5 & D \\
+ & 1 & 0 & 8 & E \\
\hline
1 & 0 & 5 & E & Y \\
\end{array}
$$

This replacement makes $E = 4$:

$$
\begin{array}{ccccc}
 & & 1 & & \leftarrow \quad \textit{carry-over} \\
 & 9 & 4 & 5 & D \\
+ & 1 & 0 & 8 & 4 \\
\hline
1 & 0 & 5 & 4 & Y \\
\end{array}
$$

Now, it can be seen that $D = 6, 7, 8,$ or 9. We can discard 8 and 9, since these digits have already been assigned. We can also discard $D = 7$, because this would make $Y = 1$ in the first column, and this digit too has been assigned. And we can reject $D = 6$, because this would make $Y = 0$ in the column, which has also been assigned. So we reject the hypothesis that $N = 5$, and go on to $N = 6$ as our next hypothesis:

$$
\begin{array}{ccccc}
 & 9 & E & 6 & D \\
+ & 1 & 0 & 8 & E \\
\hline
1 & 0 & 6 & E & Y \\
\end{array}
$$

This replacement makes E = 5:

		1	←	*carry-over*
	9	5	6	D
+	1	0	8	5
1	0	6	5	Y

This means that D = 6, 7, 8, or 9. We can discard 6, 8, and 9, because these have already been assigned. However D = 7 works perfectly, since 7 + 5 = 12. This means, of course, that Y = 2. The solution is now complete:

	9	5	6	7
+	1	0	8	5
1	0	6	5	2

Clearly, cryptarithms and alphametics are among the most challenging of all recreational mathematics puzzles. Knowledge of language structure can, of course, come in handy in solving the latter. In French, for instance, most words end in a consonant, and therefore, as in the case of English alphametics, most numerical layouts will contain different digits in the units position: e.g., 1 2 + 3 2 4 rather than, say, 1 2 + 3 2 2. However, if identical digits occur then this might suggest the plural forms of nouns, adjectives, and determiners (articles, demonstratives, etc.). Here are some examples (Antenos-Conforti, Barbeau, and Danesi 1997: 24):

D E U X	Possible solution:	7601	Meaning: "Two + six = eight"
+ S I X		+ 451	
H U I T		8052	

L E S	Possible solution:	395	Meaning: "The ten jumps"
× D I X		× 147	
S A U T S		58065	

S O N T	Possible solution:	4068	Meaning: "Are they friends?"
− I L S		− 794	
A M I S?		3274	

Many Spanish words also end in a consonant, and plural nouns, adjectives, and determiners end in an *s*. However, a large number of words end in a vowel. This means that identical digits are used in the units position more often than they are in French or English alphametics. In

Italian, on the other hand, almost all words end in a vowel. For this reason, identical digits appear frequently in the units position:

```
O T T O   Possible solution:    1771   Meaning: "Eight – six = two"
– S E I                        – 965
  D U E                          806
```

```
    L A   Possible solution:      21   Meaning: "My house"
×M I A                          × 341
C A S A                          7161
```

```
E C C O   Possible solution:    1990   Meaning: "Here is his/her hand"
    L A                            75
+ S U A                        + 465
M A N O                         2530
```

Mathematics, Magic, and Puzzles

The early histories of mathematics, magic, and puzzle-making overlap considerably. This is because the ancient magicians, mathematicians, and puzzlists were concerned with basically the same thing—unraveling hidden patterns. Indeed, no distinction was made between *numeration* and *numerology*. Numerologists translated an individual's name and birth date into numbers which, in turn, were believed to reveal the individual's character and destiny.

Numerology started with the Pythagoreans, who taught that all things were numbers and that all relationships could be expressed numerically. In Hebrew the same symbols are used for digits as for letters, and the ancient art of *gematria,* or divination, claimed that the letters of any word or name found in sacred scripture could be interpreted as digits and rearranged to form a number that contained secret messages encoded in it. The earliest recorded use of gematria was by the Babylonian king Sargon II in the eighth century B.C., who built the wall of the city of Khorsbad exactly 16,283 cubits long because this was the numerical value of his name (Clawson 1996: 48).

A thick volume could be written about the many meanings ascribed to specific numbers across the world and across history. Recall the many mystical meanings ascribed to the number 7 (chapter 1). It is found, for instance, in the Old Testament where, as part of God's instructions to Moses for priests making a blood offering, we find the following statement: "And the priest shall dip his finger in the blood, and sprinkle of the blood seven times before the Lord, before the veil of the sanctuary" (Leviticus 4:6). It is also noteworthy that God took six days to make the world and then rested on the seventh. The number 13, too, has a long history associated with mysticism. So widespread is the fear of the number 13 that it has even been assigned a name: *triskaidekaphobia*. In Christianity, 13 is linked with the Last Supper of Jesus and his twelve disciples and the fact that the thirteenth person, Judas, betrayed Jesus. Other similarly unlucky numbers exist in different parts of the world. And across cultures, people tend to think of certain things, such as dates, street addresses, or certain numbers, as having great significance. As Rucker (1987: 74) aptly phrases it, human beings seem to possess the "basic notion that the world is a magical pattern of small numbers arranged in simple patterns."

Some digits do indeed appear to have mystical qualities, even if the patterns they produce can be explained in prosaic arithmetical terms. Take, for instance, the number 9. Here are only two of the many "magical patterns" it allows us to create (Beiler 1966: 54–66):

First Pattern

$$1 \times 9 + 2 = 11$$
$$12 \times 9 + 3 = 111$$
$$123 \times 9 + 4 = 1111$$
$$1234 \times 9 + 5 = 11111$$
$$12345 \times 9 + 6 = 111111$$
$$123456 \times 9 + 7 = 1111111$$

etc.

Second Pattern

$$9 \times 9 + 7 = 88$$
$$98 \times 9 + 6 = 888$$
$$987 \times 9 + 5 = 8888$$
$$9876 \times 9 + 4 = 88888$$
$$98765 \times 9 + 3 = 888888$$
$$987654 \times 9 + 2 = 8888888$$

etc.

The fact that these patterns are generated by a specific number produces a sense of wonderment in us. Needless to say, such patterns have also been the source of many parlor-game tricks. Here is one example:

A person is asked to take some of the integers starting from 1, such as 1, 2, 3, 4, 5, and make a number with them in numerical sequence (12345). The person is then asked to multiply the number by the "magical number" 9 (12345 × 9 = 111105) and keep this product secret. Then the number wizard who asked the person to perform this multiplication asks how many integers were taken, recites a few magic spells, and, mysteriously, comes up with the product. How did the wizard do it?

The above trick is really based on the fact that 12345 × 9 + 6 = 111111, in the first pattern above. The wizard, aware of this pattern and having learned how many integers were taken at the start, knows that all he or she has to do is to take away 6 from 111111. This produces the result 111105.

.It was only after the Renaissance that numerology was relegated to the status of a pseudoscience. Paradoxically, the Renaissance at first encouraged interest in the ancient magical arts and in their relation to philosophical inquiry. Intellectuals such as Italian philosopher Giovanni Pico della Mirandola (1463–1494) rediscovered the occult roots of classical philosophy, and protoscientists such as the Swiss physician Philippus Aureolus Paracelsus (1493–1541) affirmed occult practices, partly in defiance of medieval religiosity. Both the Roman Catholic Church and the new Protestantism, however, turned sharply against magic and the occult arts in the fifteenth and sixteenth centuries. Mathematics was subsequently completely liberated from the occult mysticism in which it was shrouded in the ancient world.

But the connection between mysticism and mathematics has hardly been lost. Solving magic squares, proving a difficult theorem, or observing a mysterious manifestation of Fibonacci numbers in Nature continues to cast a magical spell over us. In fact, to this day, the boundaries between mathematics and magic are rarely clear-cut. Every mathematical idea is caught up in a system of references to other ideas, patterns, and designs that humans are inclined to dream up. And this imparts an aura of Pythagorean mysticism to that very system.

6 Puzzling Games

CHESS, CHECKERS, AND OTHER GAMES

The chess pieces are the block alphabet which shapes
thoughts; and these thoughts, although making a visual
design on the chess-board, express their beauty abstractly,
like a poem. . . . I have come to the personal conclusion
that while all artists are not chess players, all chess players
are artists.

—Marcel Duchamp (1887–1968)

Around 1400 B.C., Ramses I began construction of a temple in
the Egyptian city of Kurna, near Thebes. Three curious etchings can still
be seen on its walls today:

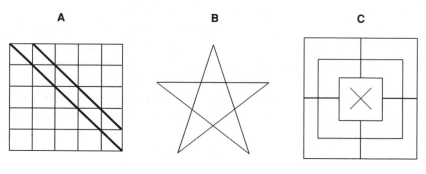

Figure 6.1

These are, more than likely, game boards that the masons who were working on the temple scratched into the stone to provide entertainment for
themselves as they toiled away (Olivastro 1993: 123). The game shown
in figure A was left unfinished, and its rules have not as yet been deci-

phered. Game B is thought to be a predecessor of *pentalpha*, a board game that is still played in Greece and other parts of the world. Markers are moved from place to place around the board. The goal is to line the markers up in a certain order or in a certain configuration, or else to remove the opponent's markers from the board according to some specified system of rules. The rules of game C have also not as yet been unraveled, but it was probably also played with markers to be lined up in some configuration.

Games B and C are based on the same principle as tic-tac-toe. The goal is to line up a specified number of markers, letters, counters, etc. in some way, as in tic-tac-toe players try to place three Xs or Os in a row, column, or diagonal of a square. The player who accomplishes this first is declared the winner.

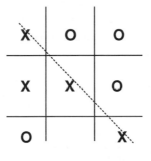

Figure 6.2

Games played according to rules and involving equipment such as boards, counters, sticks, stones, or coins have been around since the dawn of history. They are our topic on this leg of our journey through Puzzleland. Needless to say, only a handful of game genres can be discussed here. Readers can consult the many excellent books and Web sites on both the history of games and the rules for playing them (see the bibliography provided at the back of this book).

Games can be classified into four main types (Dalgety and Hordern 1999): (1) those played by manipulating objects, such as sticks, coins, or counters, with the hands (*movement and arrangement games*); (2) those played by manipulating mechanical objects, such as a Rubik's Cube (*mechanical games*), or by assembling pieces to make shapes (*assembly games*); (3) those played on a game board (*board games*); (4) those played with cards (*card games*) or dice (*dice games*). Most of these have ancient origins, and many have had implications for mathematics generally.

Movement and Arrangement Games

In cultures across the world one finds a multitude of truly inventive games involving coins, sticks, counters, and other common objects (Abraham 1933; Brooke 1963b; Beasley 1990). The following one, for instance, is found in many guises across continents and throughout the ages. The ancient Japanese called it *Hiroimono* ("things picked up"), because it is played by picking up and moving things one at a time (Costello 1996: 9). Because of its antiquity and importance to the history of games, it will be discussed in detail here:

> *There are six coins in a row on a table, three white on the left and three black on the right, with one space between the two sets. Can the arrangement of the coins be reversed by moving only one coin at a time? A coin may be moved into an adjacent empty space, or jumped over one adjacent coin into an empty space. Coins may not move backward: i.e., white coins can move only to the right and black coins to the left.*

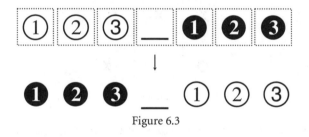

Figure 6.3

It is advisable to start by playing a simpler version of this game, in order to gain insight into its solution. So let us consider a two-coin version first. Given one white and one black coin, with a space between them, can we reverse their positions while obeying the two given rules of movement?

Figure 6.4

We can represent the white coin as W1 and the black one as B1, and the initial arrangement as a 0 (zero) state:

0. W1 — B1

We start by sliding W1 over into the empty space, thus bringing W1 and B1 next to each other:

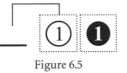

Figure 6.5

This move creates a new empty space—namely the space that W1 previously occupied. The new arrangement can be represented as follows:

1. — W1 B1

We can now jump B1 over W1 into the space, creating a new space where B1 used to be:

Figure 6.6

This new arrangement can be represented as follows:

2. B1 W1 —

For our third and final move, we can simply slide W1 over into the empty space, leaving a space where it used to be:

Figure 6.7

This is the required end state, which we were able to reach in just three moves. These are summarized below:

0. W1 — B1
1. — W1 B1
2. B1 W1 —
3. B1 — W1

Now let us turn our attention to a four-coin version of the game, whose initial arrangement looks like this:

Figure 6.8

Once again, we can show this initial state as follows:

0. W1 W2 — B1 B2

We start by sliding W2 into the empty space next to B1, creating a new space where W2 used to be:

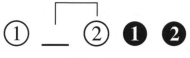

Figure 6.9

1. W1 — W2 B1 B2

We can now jump B1 over W2 into the empty space:

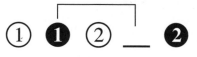

Figure 6.10

2. W1 B1 W2 — B2

Next, we can slide B2 into the empty space:

Figure 6.11

3. W1 B1 W2 B2 —

This now allows us to jump W2 over B2 into the empty space:

Figure 6.12

4. W1 B1 — B2 W2

Now, for our fifth move, we can put W1 into the empty space by jumping it over B1:

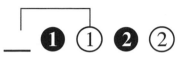

Figure 6.13

5. — B1 W1 B2 W2

This leaves an empty space, into which we can slide B1 for our sixth move:

Figure 6.14

6. B1 — W1 B2 W2

We can now jump B2 over W1 into the empty space for our seventh move:

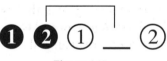

Figure 6.15

7. B1 B2 W1 — W2

For our eighth and final move, we simply slide W1 into the empty space:

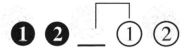

Figure 6.16

8. B1 B2 — W1 W2

The solution clearly hinges on alternating jumping and sliding moves. Using this insight, the original six-coin version of the puzzle can now be solved easily. Here is the sequence of fifteen alternating jumping and sliding moves, summarized without diagrams:

0. W1 W2 W3 — B1 B2 B3
1. W1 W2 — W3 B1 B2 B3
2. W1 W2 B1 W3 — B2 B3
3. W1 W2 B1 W3 B2 — B3
4. W1 W2 B1 — B2 W3 B3
5. W1 — B1 W2 B2 W3 B3
6. — W1 B1 W2 B2 W3 B3
7. B1 W1 — W2 B2 W3 B3
8. B1 W1 B2 W2 — W3 B3
9. B1 W1 B2 W2 B3 W3 —
10. B1 W1 B2 W2 B3 — W3
11. B1 W1 B2 — B3 W2 W3
12. B1 — B2 W1 B3 W2 W3
13. B1 B2 — W1 B3 W2 W3
14. B1 B2 B3 W1 — W2 W3
15. B1 B2 B3 — W1 W2 W3

This game is, in effect, a coin version of the basic principle embodied in Alcuin's River-Crossing Puzzle (chapter 5). Games of this type have been instrumental in the development of *critical path theory,* which studies positional relations among things. An interesting game based on "strategic arrangement" is called *Nim.* It was invented in 1901 by Charles Leonard Bouton, a professor of mathematics at Harvard University, although it resembles games that were played in ancient China (Gardner 1961b: 63–64; Agostini and De Carlo 1985: 146). In one version of the game, nine coins are arranged as shown:

Top Row →

Middle Row →

Bottom Row →

Figure 6.17

Players take turns removing one or more coins from one row. For example, a player may take one coin from the top row, or all the coins from the bottom row. The player forced to take the last coin is the loser. Now, if the first player makes the correct first move, and continues to play intelligently, that player will always win. Otherwise the opponent will always win. What is that first move? It consists in taking three coins from the bottom row. After that, any move by the first player that leaves (1) one coin in each of three rows, (2) two coins in each of two rows, (3) three coins in each of two rows, or (4) one coin in one row, two in another, and three in a third is sure to lead to a win. Readers can confirm this for themselves (see Benson 1999: 106–111 or Holt 1978: 121–122 for a general discussion).

Games such as the two discussed above involve arrangement or rearrangement. Another genre involves *orientation* instead. The following is one such game, which some puzzle historians attribute to none other than Zeno of Elea, the originator of paradoxes (chapter 4):

> Consider two coins of equal size, A and B, touching each other. If B is kept fixed and A is rolled around B's edge without slipping, how many revolutions will A have made about its own center when it is back in its original position?

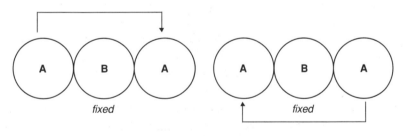

Figure 6.18

Many people come up with an incorrect solution to this puzzle. Since the circumferences of the two coins are equal, and since the circumference of A is laid out once along that of B, they argue that A must make one revolution about its own center. However, if readers actually carry out the instructions of this puzzle with, say, two quarters, they will find that the left-hand quarter will make two complete revolutions, not one.

The mathematical explanation of this apparent paradox can be found in an analysis of the figure known as a *cycloid*—a curve tracing the path traversed by a point on the circumference of a wheel as it rolls without slipping upon a straight line. The cycloid was studied and named by Italian physicist and astronomer Galileo in 1599. A point on the circumference of a circle generates a cycloid as the circle rolls forward along a line:

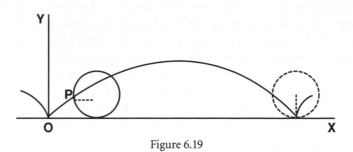

Figure 6.19

As the point P rises and falls, the wheel rotates through one complete turn. The coin game above is an enigmatological exemplification of the cycloid's properties.

The coin game below, which is quite old, involves moving components of a figure in order to alter the figure's orientation:

> *The following triangle is made up of ten coins. What is the smallest number that must be moved to make the triangle point downward?*

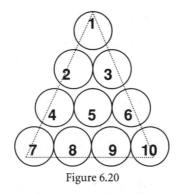

Figure 6.20

First, we move coin 1 below 8 and 9. Then we move 7 to the left of 2 and 10 to the right of 3:

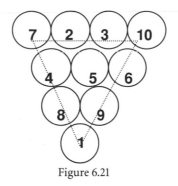

Figure 6.21

By moving these three coins we get the orientation required. The following puzzle involves a different kind of concept. It was invented by Angelo John Lewis (Professor Hoffman) in his book *Puzzles Old and New* (1893). Instead of changing the orientation of a figure, Lewis asks us to reduce the number of figures by removing certain pieces. The solution is left as an exercise for the reader (see solution 6.1):

> *Fifteen small, thin wooden rods are laid on a table, forming five identical squares. Remove three rods so as to leave only three such squares.*

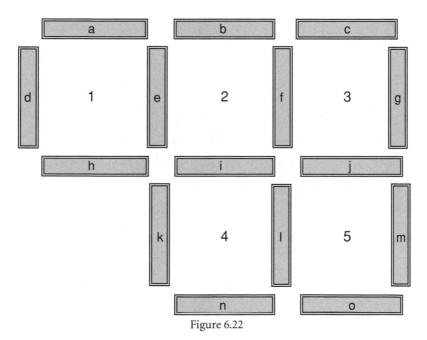

Figure 6.22

Certainly the most famous movement game of all time is the Towers of Hanoi Puzzle, devised by François Lucas in 1883 (chapter 1) after an idea appearing in the 1550 edition of Girolamo Cardano's *De Subtililate*:

> *A monastery in Hanoi has three pegs, labeled A, B, and C. A holds 64 gold discs in descending order of size—the largest at the bottom, the smallest at the top. The monks have orders from God to move all the discs to the C peg while keeping them in descending order. A larger disc must never sit on a smaller one. All three pegs can be used. When the monks move the last disk, the world will end. Why?*

To understand the mechanics of this puzzle, it is useful to start with smaller versions. To keep track of the moves, we number the discs from the top down (i.e., 1 = the smallest disc, 2 = the next larger one, 3 = the larger one after that, and so on). Let us consider a two-disc version of the game:

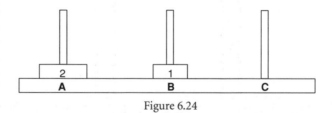

Figure 6.23

We start by moving disc 1 to peg B:

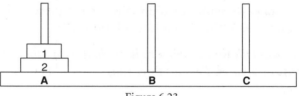

Figure 6.24

Now we can move disc 2 over to C:

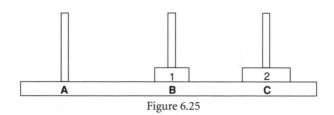

Figure 6.25

Finally, we can move disc 1 to peg C on top of 2, at which point the two discs have been transferred to C with the smaller one on top, as required:

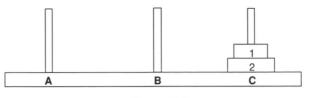

Figure 6.26

It took three moves to accomplish the task—one more than the number of disks. Now let us try the game with three discs, numbered 1, 2, and 3:

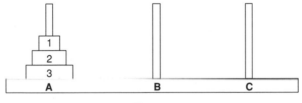

Figure 6.27

The appropriate moves (given without diagrams) are as follows:

1. Move disc 1 from A to C
2. Move disc 2 from A to B
3. Move disc 1 from C to B on top of 2
4. Move disc 3 from A to C
5. Move disc 1 from B to A
6. Move disc 2 from B to C on top of 3
7. Move disc 1 from A to C on top of 2, which is itself on top of 3

This time it took seven moves to accomplish the task—one more than twice the number of disks. If you play the Towers of Hanoi game with four, five, and higher numbers of discs, you will find that the number of moves needed to transpose n discs according to the given conditions is $2^n - 1$. In the case of 2 discs ($n = 2$), the number of moves, as we saw above, is $2^2 - 1 = 3$; in the case of 3 discs ($n = 3$), the number of moves, as we also saw above, is $2^3 - 1 = 7$. In Lucas's puzzle the number of moves required is, therefore, $2^{64} - 1$, which, at one move per second, amounts to 5.82×10^{11} years. It would thus take nearly six billion centuries to complete the puzzle, even assuming the monks made no errors along the way!

One of the most ancient of all arrangement games is the Josephus Puzzle (the version here is adapted from Kasner and Newman 1940: 173–174). It postulates a ship carrying a number of people, some of whom must be thrown overboard to prevent the ship from sinking. The victims must be chosen according to a numerical rule or pattern, but we wish to arrange matters so as to preserve a certain subset of the passengers at the expense of the others. For instance, suppose the ship is carrying thirty people, fifteen As and fifteen Bs, and we wish to sacrifice the As to preserve the lives of the Bs. One way to do this is to place the passengers in a circle as shown below and announce that every ninth person is to be thrown overboard. Counting starts at the arrow and proceeds clockwise around the circle (which shrinks as the hapless As go over the side):

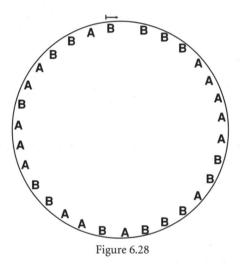

Figure 6.28

Versions of the Josephus Puzzle are found in many cultures in different eras. Elaborate versions of this puzzle were devised and studied by famous mathematicians, including Leonhard Euler.

Mechanical and Assembly Games

Mechanical games that involve opening a device, finding a secret compartment, or disassembling an object became highly popular among the merchant classes during the Middle Ages. Girolamo Cardano described some of these in his 1557 treatise *De Rerum Varietate*. But only after the invention of sheet-metal stamping, around 1850, did large-scale manufacturing of such games become possible. By the end of the nineteenth century objects composed of various parts and involving some form of manipulation were highly popular as toys and as educational games.

One of these—Sam Loyd's 14/15 Puzzle, which he devised in 1878—became an international craze (chapter 1). Loyd's gadget was made up of 15 consecutively numbered sliding blocks, placed in a square tray large enough to hold 16 such blocks. The blocks were arranged in numerical sequence except for 14 and 15, which were installed in reverse order—hence the puzzle's name. The game challenged solvers to get the blocks into numerical sequence from 1 to 15, by sliding them, one at a time, into an empty square:

1	2	3	4
5	6	7	8
9	10	11	12
13	15	14	

empty square

Figure 6.29

The puzzle, as it turns out, is impossible to solve, but hordes of people attempted to do so just the same. As this episode in the history of puzzles shows, people simply cannot ignore a challenge, no matter what the costs are in time and energy. Such is the sway the puzzle instinct has over the imagination. So popular was the 14/15 Puzzle in America that employers posted notices prohibiting employees from playing the game during work hours. In Germany and France, it was denounced by newspapers as a greater scourge than alcohol or tobacco.

In mathematical terms, Loyd's puzzle cannot be solved because its arrangement does not constitute a *Hamiltonian circuit,* named after the Irish mathematician William Rowan Hamilton (1805–1865). The demonstration of this is long and complicated, and thus beyond the scope of this book. A clear and detailed proof can be found in Benson 1999 (74–87). Lucas discussed Loyd's game at length in his *Mathematical Recreations.* He then proceeded to construct his own solvable version of it to appease a frustrated public of puzzle addicts. In Lucas's gadget, there are four moveable blocks, A, B, C, and D. Around the rim of the tray are empty spaces into which the blocks can be moved:

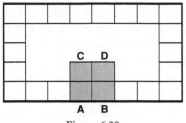

Figure 6.30

Lucas's game requires us to reverse C D to D C and A B to B A. A B can be reversed by moving B one square to the right, and then moving A around the rim clockwise, as shown, into the space to the right of B:

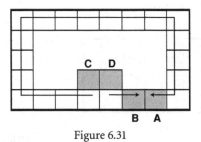

Figure 6.31

Now C and D can be interchanged by moving them as shown:

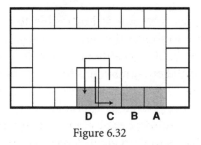

Figure 6.32

In 1921 an inventor named P. A. MacMahon came up with a very popular device called the Thirty Colored Cubes Puzzle, which was, in all likelihood, the prototype for the Rubik's Cube, invented by the Hungarian Ernö Rubik in 1975. Rubik, a professor at the Budapest School of Applied Arts, claims to have invented it as an exercise in spatial reasoning for his students. By 1982 it had become an international craze, with ten million sold in Hungary alone (more than the population of the country). It is estimated that by the end of the twentieth century over one hundred million Rubik's Cubes had been sold worldwide. The puzzle is made up of smaller, colored cubes, so that each of the six faces of the large cube is a different color. The colors can be scrambled by twisting sections of the cube around any axis. The goal is to scramble the colors and then return the cube to its initial configuration. There are 43,252,003,274,489,856,000 possible arrangements of the small cubes, only one of which is the initial (and target) one (Frey and Singmaster 1982; Rubik 1987)! Solving the cube can be a mind-numbing task, yet people keep at it, literally night and day, as the enduring popularity of the puzzle indicates. Incidentally, a man named Larry Nichols patented a similar puzzle cube in 1957 (Costello 1996: 148). In 1984, Nichols won a patent infringement lawsuit against the Ideal Toy Company, the makers of Rubik's Cube.

Cube puzzles have also piqued the interest and stimulated the imagination of artists. The metal puzzle sculptures of the Spaniard Miguel Berrocal (b. 1933), for instance, challenge the viewer to take them apart and then put them together again. His *Cofanetto* sculpture, which is a tribute to the Romeo and Juliet story, can be disassembled into eighty-four pieces and reassembled to make two persons.

One of the most popular arrangement puzzles in history is the *jigsaw puzzle,* which is solved by assembling interlocking pieces to form a picture or figure. It was invented by British mapmaker John Spilsbury around 1760 as a toy to educate children about geography (Hannas 1972, 1981). Jigsaw puzzles for adults were put on the market around 1900. Most of them had no guide or picture on the box (unlike children's versions of the puzzles). Parker Brothers introduced picture guides in 1908. This new form of the puzzle became so successful that in 1909 the company devoted its entire factory production to it. Made of wood, the Parker Brothers jigsaw puzzles were extremely expensive—a 500-piece puzzle cost $5 in 1908, when the average wage was $50 per month. By 1933, they were more and more often made of cardboard and were much more affordable, propelling sales to nearly ten million per week in just the United States. Retail stores offered free puzzles with the purchase of certain gadgets, and a twenty-five-cent jigsaw puzzle in a box, called the *Jigsaw Puzzle of the Week,* was sold at newsstands every Wednesday. Today, the jigsaw puzzle remains one of the more popular types of games, enjoyed by children and adults alike. Specialty stores throughout North America sell jigsaw puzzles to suit all tastes.

Assembly puzzles attracted the interest of several mathematicians in the twentieth century. In 1957, for instance, Solomon W. Golomb invented the game of *polyominoes.* A polyomino is a two-dimensional shape formed by joining square, flat pieces of paper or plastic. There are several basic types of polyominoes: the *domino* is a two-square figure, the *tromino* a three-square figure, the *tetromino* a four-square figure, and the *pentomino* a five-square figure. They can be assembled into a variety of forms. For instance, trominoes can be assembled to produce the letters T and L as follows:

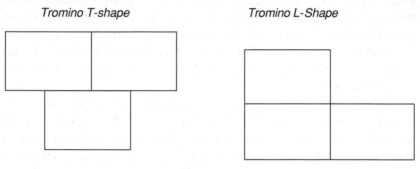

Figure 6.33

The following shows the high degree of imaginative thinking that is required to solve polyomino puzzles (adapted from Golomb 1965: 26):

> *Here are twelve figures, each composed of five identical squares placed edge to edge. You will notice that one figure resembles a V, and another an X. How can enlarged copies of each of these two figures, three times as wide and three times as tall as the originals, be assembled, with each copy using nine of the original figures?*

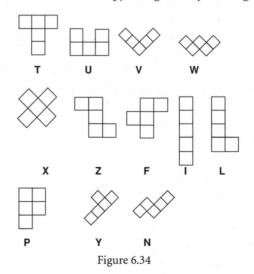

Figure 6.34

The solution is truly difficult to envision. It is left as an exercise for readers, who may want to cut out cardboard versions of the figures above and attempt to solve the puzzle with them. (For Golomb's answer, see solution 6.2.)

In 1958, the Danish poet Piet Hein (b. 1905) invented a three-dimensional elaboration of Golomb's two-dimensional game, which he called Soma cubes (Gardner 1998: 70–71). As with polyominoes, the goal of Hein's game is to join the cubes together to form shapes. Here is an example of a skyscraper that can be made with Soma cubes:

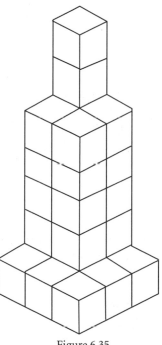

Figure 6.35

Hein is also known for having invented the game of Hex in 1942. This is played on a diamond-shaped board made up of interlocking hexagons (hence the abbreviated name Hex). A typical board has eleven hexagons on each edge. One player has a supply of black counters, the other an equal number of white. The players take turns putting one of their counters on any unoccupied hexagon. The object is to complete a continuous chain of counters from one edge of the board to the opposite edge. Players try to block each other's attempts to construct a chain, as they try to complete their own.

In 1974, the British physicist Roger Penrose invented an intriguing kind of assembly game, which has come to be called, appropriately, the Penrose Tile Game. It involves making *nonperiodic tilings* of the plane with two figures. A *periodic tiling* is a design that recurs horizontally or vertically across the plane. The grid design of a sheet of graph paper is an example of such a tiling, since the same pattern of white square after white square recurs both vertically and horizontally. A *nonperiodic tiling* does not involve a recurring pattern. For example, the following tiling, which divides the plane into right triangles, is nonperiodic:

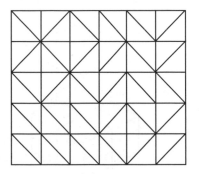

Figure 6.36

The Penrose Tile Game produces an infinite number of different non-periodic tilings of a surface, using two figures called a *dart* and a *kite*. Penrose used the golden ratio φ (chapter 3) to design his darts and kites:

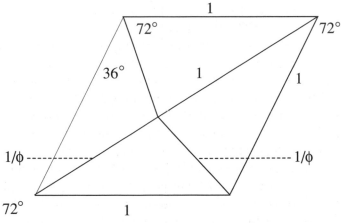

Figure 6.37

This figure can be split into a dart and a kite as follows:

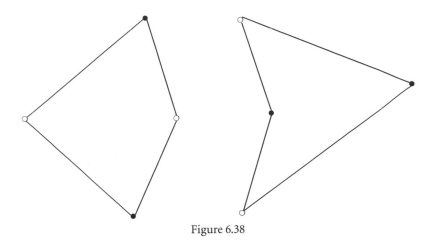

Figure 6.38

As many kites and darts as desired can be made. The goal of the game is to join them to produce different kinds of figures, such as suns, stars, etc. This is left as an exercise for the reader. In a basic sense, Penrose's tiles, like Golomb's polyominoes or Hein's Soma Cubes, are works of pure art, inviting the eye to detect pattern and symmetry for their own sake. They produce the same kind of aesthetic pleasure that comes from looking at a cubist painting.

Board Games

Myriads of board games have been invented throughout the ages, from the ancient tic-tac-toe games scratched on the walls of the temple at Kurna to more contemporary games such as chess, backgammon, checkers, Monopoly, Parcheesi, Concentration, Jeopardy, and Scrabble. One of the oldest known board games was found at Ur, a city in ancient Sumer. Archaeologists believe the game is about 4,500 years old. It was probably a race game, in which players threw dice and moved pieces around a track. Another ancient game, known in the West as Go, originated more than 2,000 years ago in Japan, where it is called *weiqi*. Go is played on a board with nineteen horizontal and nineteen vertical lines. The object of the game is to capture a territory by surrounding it with black and white pieces called stones. Today, professional Go players earn large salaries in many Asian countries.

The so-called *mancala* games—ancient African and Asian games involving pebbles, seeds, or other small objects that must be scooped out of a board—are the prototypes for modern day *role-playing* games, such as Dungeons & Dragons, invented in the 1970s. In the latter game—which, of course, is not a board game—one player takes the role of "gamemaster." The other players assume the roles of various characters with certain attributes, such as strength or magical ability. Players lead their characters through imaginary adventures, such as finding treasures in dungeons guarded by monsters. The gamemaster tells the players what happens to them through each stage of the adventure.

Perhaps the most popular board game of all time is chess, in any and all of its different cultural versions—e.g., *shogi* in Japan and *xiang qi* in China (Bell and Cornelius 1988). Chess is played using specially designed pieces on a board of sixty-four alternating light and dark squares in eight rows of eight squares each. The vertical columns on the board that extend from one player to the other are called *files,* and the horizontal rows are called *ranks.* The diagonal lines across the board are called *diagonals.* Each player controls an army composed of eight pawns and eight more powerful pieces: one king, one queen, two rooks (sometimes called castles), two bishops, and two knights. The two armies are of contrasting colors, one light and the other dark, and are always called "white" and "black" regardless of their actual colors:

Chessboard with pieces laid out
Figure 6.39

White always moves first, and the players then alternate turns. A move consists of transferring a piece to another square that is either vacant or occupied by an opponent's piece. If it is occupied, the opponent's piece is captured (i.e., removed from the board and replaced by the capturing piece). The only exception is the king, which is never captured. A move to capture is not required unless it is the only possible move.

Only one piece may be moved at each turn except when *castling*—a special maneuver involving the king and one rook. The king may be moved one square in any direction, but not to any square where an opposing piece could capture it on the next move. The queen may be moved as far as desired in any uninterrupted direction: horizontally, vertically, or diagonally. The rook may be moved as far as desired either horizontally or vertically. The bishop may be moved as far as desired diagonally, so that each bishop remains on squares of the same color throughout the game. The knight is the only piece that does not normally move in a straight line, and also the only piece that can move over or around other pieces in its way (except that castling moves the king and rook past each other). The knight moves two squares horizontally or vertically and then one square at a right angle to the first part of its path, so that it always ends on a different-colored square than it began on. Each pawn, on its first move only, may be moved straight ahead either one or two squares to a vacant square. After that it may be advanced only one square at a time. Pawns, unlike the other pieces, do not capture in the direction they are moved, but diagonally one square forward; this is the only time a pawn may move diagonally. If a pawn is advanced two squares on its first move and lands next to an opponent's, the opponent's pawn may capture it as if it had advanced only one square. When a pawn reaches the last rank on the opposing side of the board, it is promoted—i.e., converted to any other piece of the same color (except another pawn or the king).

The object of the game is to trap the opposing king so that it cannot escape being captured (the game is not played through to the actual capture). A king that could be captured by an opposing piece in one move is said to be *in check*. Check does not have to be announced, but the player whose king is in check must attempt to escape on the next move. There are three ways of doing so: (1) moving the king to a safe square; (2) capturing the attacking piece; and (3) cutting off the attack by interposing a piece between the attacking piece and the king. If none of these is possible, the king is *checkmated* (from the ancient Persian *shah mat*, meaning "the king is dead") and the player who brings about the checkmate wins.

Chess is descended from *chaturanga*, a game popular in India in the sixth century A.D. By the end of the tenth century, chess was well known throughout Europe, attracting the fanatical interest of rulers, philosophers, and poets (Falkener 1892). A 1474 translation by the English printer William Caxton (c. 1421–1491) of a chess treatise written by a Dominican friar standardized the rules of chess for a period of time. But it was not until international competition developed in the subsequent century that the rules were regularized once and for all. The game was popularized in the eighteenth century by a Syrian named Philip Stamma, acclaimed as the pioneer of modern chess technique, and the French chess master François-André Danican Philidor, who published the influential treatise *Analysis of the Game of Chess.*

During the early 1990s, computer scientists at IBM developed a chess computer named Deep Blue that was capable of analyzing millions of chess positions every second. In 1996, world chess champion Garry Kasparov defeated the computer in a highly publicized match, 4 games to 2. He faced an improved version of Deep Blue a year later in a rematch. The enhanced computer was capable of processing two hundred million positions per second. Kasparov won the first game of the rematch, but after Deep Blue secured draws in games 3, 4, and 5 and victories in games 2 and 6, Kasparov was declared the loser, 2.5 games to 3.5. The event marked the first-ever defeat of a world chess champion by a computer.

Puzzles based on chess were already popular during the twelfth and thirteenth centuries. The Arabic mathematician Ibn Kallikan, for instance, used the chess board in 1256 to pose his famous Grain Problem, in which each square contains twice as many grains of wheat as the previous one. As we saw in chapter 1, on the sixty-fourth square there would be 2^{63} grains of wheat.

Sam Loyd was a prolific inventor of chess puzzles. The following one appeared in March 1867 (the version here is paraphrased from White 1913: 42):

> Place the queen on a chessboard and pass her over the entire sixty-four squares and back again to her starting point in just fourteen moves.

To solve this puzzle, start the queen in a corner square and proceed as shown below:

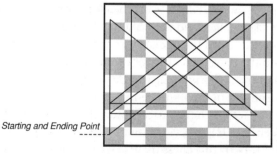

Starting and Ending Point

Figure 6.40

Raymond Smullyan's 1979 *The Chess Mysteries of Sherlock Holmes* is a classic compilation of chess puzzles. Smullyan shows that chess is essentially a board-game version of deductive puzzles (chapter 4). And, appropriately enough, he presents his chess puzzles through a series of dialogues between master fictional sleuth Sherlock Holmes and his assistant Watson. Each puzzle requires the solver to deduce certain events in a game's past: e.g., *On what square was the white queen captured? Is the white queen the original queen or a promoted pawn?* etc.

A board game that may be older than chess is checkers. Like chess, it is played by two people on a checkered board of sixty-four alternately colored squares. Predecessors of the game have been traced to 1600 B.C. in ancient Egypt. Many game historians, however, trace the origin of the modern form of the game to southern France around the twelfth century. The earliest modern account of checkers appears in a book written by a certain Antonio Torquemada and published in Valencia, Spain, in 1547.

Checkers is played by placing the pieces, twelve to each player, on the dark squares. They may be moved only diagonally forward, one square at a time. A piece, or checker, may not be moved to a square occupied by another, but may jump over an opposing piece to a vacant square immediately beyond. Such a jump "takes" the opposing piece, which is removed from the board. If another opposing piece is diagonally adjacent to the new position, with another empty space beyond, the second piece may also be jumped and taken; jumps may continue as long as there are opposing pieces to be jumped and empty spaces to land in. When a piece on either side has been advanced to the last row on the opposite side, it is *crowned*—that is, made a *king*. A second checker is placed on the crowned one to identify it. Like other pieces, kings can be moved only diagonally, one square at a time, except that jumps may go on as long as possible. Kings, however, may be moved either backward or forward, an additional power that bestows a great advantage to the player who has the most kings. The object of the game is to clear the board of the opponent's checkers or to block the opponent's checkers so that they cannot be moved.

Chess and checkers are constantly being compared, with chess generally agreed to be harder to play. But, as Edgar Allan Poe argues in the opening of his "Murders in the Rue Morgue," this is a misconception:

> I will, therefore, take occasion to assert that the higher powers of the reflective intellect are more decidedly and more usefully tasked by the unostentatious game of draughts [checkers] than by all the elaborate frivolity of chess. In this latter, where the pieces

have different and *bizarre* motions, with various and variable values, what is only complex is mistaken (a not unusual error) for what is profound. The *attention* is here called powerfully into play. If it flag for an instant, an oversight is committed, resulting in injury or defeat. The possible moves being not only manifold but involute, the chances of such oversights are multiplied; and in nine cases out of ten it is the more concentrative rather than the more acute player who conquers. In draughts, on the contrary, where the moves are *unique* and have but little variation, the probabilities of inadvertence are diminished, and the mere attention being left comparatively unemployed, what advantages are obtained by either party are obtained by superior *acumen*. (Poe 1960: 50)

As with chess, many puzzles based on checkers have been devised. One that finds its way into virtually every puzzle anthology because of its apparent difficulty, yet which has a deceptively simple solution, is the following:

> *If two opposite corners of a checkerboard are removed, can the checkerboard be covered by dominoes? Assume that the size of each domino is the size of two adjacent squares of the checkerboard. The dominoes cannot be placed on top of each other and must lie flat.*

Domino

Board with two white opposite corners removed

Figure 6.41

It is not possible to cover the altered checkerboard in this way, for the simple reason that the two squares that are removed are of the same color. A domino placed on the checkerboard always covers a white and a black square. With two opposite white corners removed, the board does not have an equal number of black and white squares for dominoes to cover.

Another classic puzzle in this genre is the following. The solution is, again, remarkable for its simplicity (see solution 6.3):

> Place eight checkers on a checkerboard so that no two checkers lie in the same row, column, or diagonal.

Chess and checkers are often perceived to be metaphors for life. In Swedish director Ingmar Bergman's (b. 1918) apocalyptic 1956 movie *The Seventh Seal*, the narrative revolves around a game of chess between a knight and Death. The message of the movie is obvious—if somehow we were to understand the game of life with our intellect (= puzzle instinct?), then we could win the greatest game of all, cheating Death. But as it turns out, this is a hollow dream—in the movie and in life.

Card and Dice Games

The first playing cards in Europe date from the medieval period. Known as *tarot cards,* they were used for fortune-telling. They were probably introduced into Europe either by the crusaders, between 1095 and 1270, or by the Roma (Gypsies) in fourteenth-century Italy. A full deck consists of 78 cards: the *minor arcana* (56 suit cards) and the *major arcana,* also known as *trumps* (22 pictorial symbol cards). The minor arcana consists of suits of wands, cups, swords, and pentacles. There are 14 cards for each suit: 4 court cards (king, queen, knight, and page) and cards numbered from ace to 10. The major arcana consists of a Fool (also called a Madman) and pictorial cards numbered from 1 to 21.

Many of the earliest tarot decks were designed by artists, such as the German artist Albrecht Dürer, who was fascinated, as we saw in the previous chapter, by the magical qualities of numbers. The cards symbolize natural forces and human virtues and vices. Fortunes are determined by interpreting the combinations formed as the cards are dealt out.

The many card decks that we use today are direct descendants of the tarot cards. In the standard deck, there are 13 cards for each of the 4 suits, consisting of 3 face cards (king, queen, and jack), and cards numbered from ace to 10. In addition to these, one or two cards are known as jokers (corresponding to the tarot Fool). Other changes in the standard deck have been few. Indices—the small suit symbols at opposite corners of the cards—were added in the late nineteenth century.

Puzzles based on cards have been instrumental in the development of *probability theory*. Here is an example of a simple puzzle in this genre:

> *How many different ways can four cards be drawn from a standard deck?*

The answer is $52 \times 51 \times 50 \times 49 = 6,497,400$. The reasoning behind the solution goes like this. Any one of the 52 cards can be drawn first, of course. Each of the 52 possible first cards can be followed by any of the remaining 51 cards, drawn second. Since there are 51 possible second draws for each possible first draw, there are 52×51 possible ways to draw two cards. Now, for each draw of two cards, there are 50 cards left in the deck that could be drawn third. Altogether, there are $52 \times 51 \times 50$ possible ways to draw three cards. Reasoning the same way, it is obvious that there are $52 \times 51 \times 50 \times 49$, or 6,497,400 possible ways to draw four cards from a standard deck:

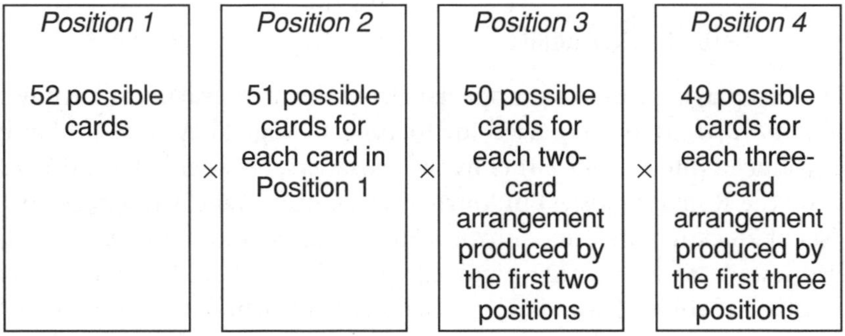

Position 1		Position 2		Position 3		Position 4
52 possible cards	$\times$	51 possible cards for each card in Position 1	$\times$	50 possible cards for each two-card arrangement produced by the first two positions	$\times$	49 possible cards for each three-card arrangement produced by the first three positions

Figure 6.42

Now, to find the chances of drawing four aces in a row, it is necessary to determine, first, the number of four-ace draws there are among the 6,497,400 possible draws. We start by looking at each outcome, draw by draw. There are four aces that can be drawn first. Now, for each one of these, there are three remaining aces that can be drawn second. Then, for each two-ace draw, there are two aces that could be drawn third. Finally, after three aces have been drawn, only one remains. So the total number of four-ace arrangements is $4 \times 3 \times 2 \times 1 = 24$. Thus, among the 6,497,400 ways to draw four cards, there are 24 ways to draw four aces. The probability of doing so is, therefore, $24/6,497,400 = .0000036$, which makes it highly unlikely.

Such puzzles put on display the relation between actual outcomes and hypothetical thinking—a form of thinking that underlies most of scientific reasoning. Take, as another example, the following classic puzzle (Kasner and Newman 1940: 243–244):

> *Since there are four aces in a deck, the probability of drawing an ace from 52 cards is 4/52 = 1/13. But what is the probability of drawing either an ace or a king from the deck in one draw?*

This is an example of the probability of *mutually exclusive* or *alternative events* (see solution 6.4).

Dice puzzles, too, have been used to illustrate various aspects of probability theory. The following is a classic in this genre (Kasner and Newman 1940: 244):

> *What is the probability of obtaining either a 6 or a 7 in throwing a pair of dice?*

We can start by listing the number of ways to throw either a 6 or a 7:

Outcome: 6		Outcome: 7	
First Die	Second Die	First Die	Second Die
1	5	1	6
2	4	2	5
3	3	3	4
4	2	4	3
5	1	5	2
		6	1

Now, there are 36 possible throws of two dice, because each of the 6 faces of the first die is matched with any of the 6 faces of the second one. Of these 36 possible throws, 11 produce either a 6 or a 7 (as the table above shows). Therefore the probability of throwing either a 6 or a 7 is $^{11}/_{36}$.

The foundations of modern-day probability theory were laid by Girolamo Cardano, himself an avid gambler, in the sixteenth century. Cardano was the first to discuss and calculate the probability of throwing certain numbers and of pulling aces from decks of cards. Cardano presented his results in his *Book of Games of Chance*, discussing the likelihood of winning fair games, as well as suggesting ways to cheat! In the subsequent century the French mathematicians Blaise Pascal (1623–1662) and Pierre de Fermat developed Cardano's ideas into a branch of mathematics.

The whole idea of taming chance by "mathematizing" it into a theory reveals a desire to conquer uncertainty. But, as the French writer François de La Rochefoucauld (1613–1680) argued in his *Maxims* (1665), this desire is foolish. Life is not predictable or predestined, let alone controllable. One's final destiny, La Rochefoucauld emphasized, will always constitute a disquieting psychic problem, no matter how many ingenious mathematical artifacts we create to control it. Probability theory, which is essentially a means of assigning mathematical pattern to chance, is, La Rochefoucauld suggested, an intellectual oxymoron.

The Game of Life

Stories about games fill the pages of human history. Today there is no toy or hobby shelf without an assortment of mechanical, board, card, and computer games, among other kinds. Game shows on television attract a large audience. And game tournaments are among the most popular of all human activities.

All this strongly suggests that, at an unconscious level, we perceive life as a kind of existential game. As mentioned above, this was captured magnificently by Bergman in *The Seventh Seal*. It has also been captured in puzzle form by the brilliant British mathematician John Conway, the inventor of a board game that he calls, appropriately enough, Life (see Rucker 1987: 111–119). The board is divided into squares, each of which may be filled with a marker. Every square has eight neighboring squares, four sharing the square's sides and four diagonally adjacent, so each marker may have from 0 to 8 neighbors:

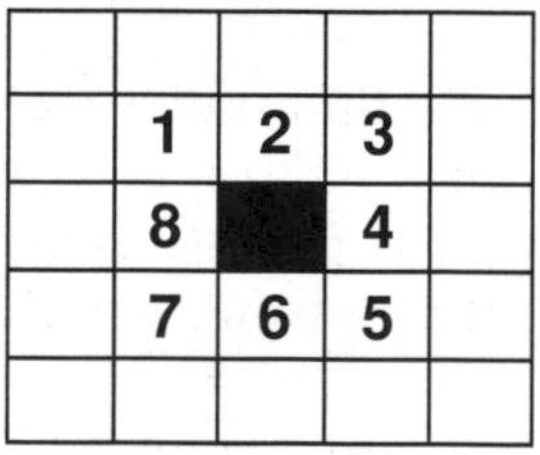

Figure 6.43

The original distribution of markers is called the *first generation*. Each generation gives rise to the next according to the following three rules:

(1) Every marker that has 2 or 3 neighbors stays on the board and continues to the next generation.
(2) Every marker with 4 or more neighbors is removed from the board in the next generation, as are those with 0 or 1 neighbor. Such markers are said to *die out*.

(3) Each empty square with exactly 3 neighbors with markers on them has a marker placed on it in the next generation. The square is said to *come to life.*

These rules apply concurrently. Some initial configurations die out, others survive, changing and reproducing themselves. The metaphorical association between Conway's game and human life is transparent. The game, like life itself, unfolds mysteriously as sequences of reproducing patterns. Here is an example of a pattern reproducing itself on a 6 × 6 board (from Paulos 1991: 212):

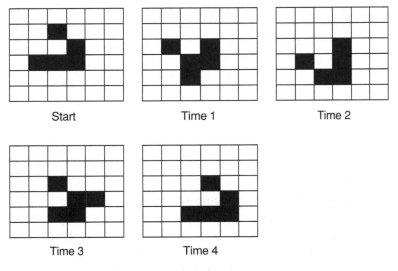

Start   Time 1   Time 2

Time 3   Time 4

Figure 6.44

Searching for meaning in the real Game of Life is akin to finding the rules established by some master of the game—a theme captured brilliantly by the German novelist and poet Hermann Hesse (1877–1962) in his last novel, *Das Glasperlenspiel* (1943, translated as *Magister Ludi,* 1949). In that novel, the meaning of life is revealed gradually to Magister Ludi in the small-scale experiences he derives from playing an eternal bead game that involves repeating patterns, not unlike those found in Conway's game.

7 The Puzzle of Life

The fact is that our lives are largely spent in solving
puzzles; for what is a puzzle but a perplexing question?
And from our childhood upwards we are perpetually
asking questions or trying to answer them.
　　　　　　　　　—Henry E. Dudeney (1857–1930)

Above all else, the purpose of this excursion into Puzzleland
has been to argue, primarily by illustration, how creative and engaging
the puzzle instinct is. Originating at the dawn of civilization, puzzles
are among the oldest products of human ingenuity. The great American
writer Henry Miller (1891–1980) once proclaimed that many of the seemingly trivial things that we have produced throughout history may exist
to counteract a deeply rooted feeling within us that life may have no
meaning. That is certainly true of puzzles. In their own miniature way,
and as trivial as they may seem, puzzles fill an existential void, so to speak,
that we would otherwise feel constantly within us, by providing small-
scale experiences of the large-scale questions that life poses. The puzzle
instinct is, arguably, as intrinsic to human nature as are humor, language,
art, music, and all the other creative faculties that distinguish humanity
from other species.

In this final chapter, a synoptic overview is obviously in order, especially of the four themes that have guided us on our journey through
Puzzleland. The first one is that puzzles stimulate the imagination and
are solved primarily by a form of insight thinking that the philosopher
Charles Peirce called *logica utens*. The second is that puzzles have, seren-

dipitously, often led to discoveries in mathematics. Zeno's paradoxes led to the invention of the calculus; Alcuin's River-Crossing Puzzle (chapter 5) prefigured modern-day critical path theory; Euler's Königsberg's Bridges Puzzle led to the development of network theory and topology; and the list could go on and on. The third theme is that puzzles are pleasurable in themselves, as small works of art, producing what may be called an aesthetics of mind. In this way, they are similar to humor. Indeed, puzzles and humor can be seen as two sides of the same aesthetic coin. The fourth theme is that many puzzles emerged in antiquity in tandem with the magical and occult arts, suggesting that puzzle-making was intertwined originally with mysticism and, thus, with the inexorable search for meaning in life.

Puzzles and the Imagination

The first theme that has guided us on our trek through Puzzleland is that puzzles are solved primarily by a form of insight thinking that the philosopher Charles Peirce called *logica utens*. This inheres in making guesses, trying out hunches on the basis of previous experience, and developing relevant insights from them. This has always been the hallmark of all the great puzzles, not to mention of mathematical invention in general. Consider, as a case in point, Ahmes' estimation of the value of π—the ratio of the circumference of a circle to its diameter, which we know today to be approximately 22/7 or, to five decimal places, 3.14159 (Beckmann 1971; Blatner 1997). The relevant puzzle in the Rhind Papyrus is Problem 48:

> *What is the area of a circle inscribed in a square that is 9 units on its side?*

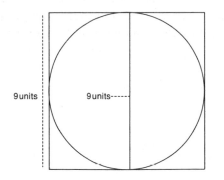

Figure 7.1

Attempting to solve this puzzle with a purely deductive approach—which Ahmes called accurate reckoning—leads absolutely nowhere. So the clever Ahmes (or whoever the real author of the Rhind Papyrus was) solved it instead with an *Aha!* insight he must have attained by contemplating the figure. *What if the circle is transformed into a polygon?* He proceeded to do exactly that by trisecting each side of the square, as shown in the diagram below, thus producing nine smaller squares within it (each 3 × 3). He then drew the diagonals in the corner squares, as shown. Such modifications to the diagram produce an octagon, which Ahmes assumed to be close enough in area to the circle for the practical purposes of his puzzle:

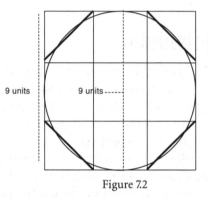

Figure 7.2

The area of the octagon is, clearly, equal to the areas of the five inner squares (which form the outline of a cross) plus half the areas of the four corner squares (= the area of two squares). Its area is thus equal to the sum of the areas of seven small squares. The area of one small square is, of course, 3 × 3, or 9 square units. The total area of seven such squares is, therefore, 9 × 7, or 63 square units. With a bit of convenient cheating, the resourceful Ahmes assumed that the circle's area was just a bit larger than that of the octagon—namely, 64 (= 8^2) square units. Ahmes then estimated the value of π as follows:

Area of circle:		$\pi r^2 = 64$
diameter:		9
radius (r):		9/2
So r^2	=	$(9/2)^2$
	=	20.25

Thus, since $\pi r^2 = 64$, and $r^2 = 20.25$:

π	=	64/20.25
	=	3.16049 . . .

A strikingly similar insight was used over one thousand years later by Archimedes, as mentioned briefly in chapter 4, who inscribed a polygon with 96 sides in a circle to calculate the value of π as somewhere between $3^{10}/_{71}$ and $3^{1}/_{7}$. In A.D. 264, the Chinese mathematician Liu Hui inscribed a polygon of 3072 sides to calculate π to the fifth decimal place.

The calculation of π has not been a trifling matter in the history of human civilization. A world in which π is not known is, of course, conceivable. But what we now know about circular objects in the world would be much more rudimentary. As Kasner and Newman (1940: 89) aptly put it, "our ability to describe all natural phenomena, physical, biological, chemical or statistical, would be reduced to primitive dimensions."

As this example shows, puzzle-solving is hardly ever a matter of straightforward accurate reckoning. As Ahmes himself demonstrated in most of the solutions he provided in the Rhind Papyrus, it almost always involves insight thinking in tandem with accurate reckoning. The source of such thinking is, of course, the imagination, the faculty of mind that allows us to see with the mind's eye what is not instantly obvious. As Dehaene (1997: 151) puts it, it is what produces the "illuminations" that mathematicians claim to see within their minds:

> They say that in their most creative moments, which some describe as "illuminations," they do not reason voluntarily, nor think in words, nor perform long formal calculations. Mathematical truth descends upon them, sometimes even during sleep.

Insight thinking is an intrinsic attribute of all human beings, no matter where they live, what intelligence they possess, or what they have experienced. Accurate reckoning, on the other hand, is something that must be acquired, nurtured, and rehearsed. It constitutes what Peirce called *logica docens*. And, for some strange reason, it can sometimes even constitute an obstacle to insight thinking. Consider the following puzzle:

> *A string is wound in a symmetrical spiral around a rod, from one end to the other. The string goes exactly four times around the rod. The circumference of the rod is 4 cm. and its length is 12 cm. Find the length of the string.*

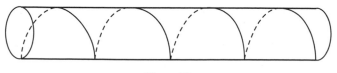

Figure 7.3

To some people this puzzle appears, as stated, to be unsolvable. Their difficulty is caused, arguably, by having solved too many problems in geometry mechanically, an experience that tends to condition people to look only for what is routine, rather than what is exceptional. The insight crucial to solving this problem comes from thinking about the rod as if it were, say, a hollow tube. What would happen if the tube were slit down one side (at the point on its rim where the string begins and ends) and flattened? Let us do exactly that and see what this insight yields:

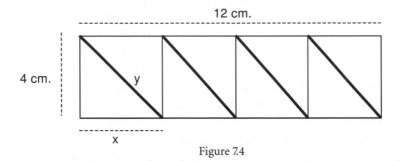

Figure 7.4

This figure now shows all we need to know to solve the problem. Our insight turned a seemingly intractable problem into a routine one. The rest of the solution is, in fact, mechanical and rather uninteresting. It proceeds as follows. We let x be the length shown above—the length of the rod over which the string made one complete turn around it. Since the string goes exactly four times around the rod, then x is ¼ of 12 cm., or 3 cm. Letting y stand for the length of that portion of the string shown, and using the Pythagorean Theorem, it is obvious that $y^2 = 3^2 + 4^2$ and, thus, that $y = 5$. Since the four portions of the string are equal, the total length of the string is $4 \times 5 = 20$ cm. The point to be emphasized here is that no such calculations would have been possible without the initial insight. Significantly, that insight can now guide us in solving future problems of the same kind. Insights are rarely forgotten, becoming almost invariably a part of memory.

Take, as one last example of pure insight thinking, a famous puzzle devised by Don Lemon (real name, Eli Lemon Sheldon) in his 1890 *Everybody's Book of Illustrated Puzzles* (adapted from Costello 1996: 11):

> *Five boarders live in a house with a garden. The owner of the house wants to divide the garden among the five. There are ten trees in the garden, laid out in a particular way. How can the owner divide it so that each of the five boarders receives an equal share of the garden and two trees?*

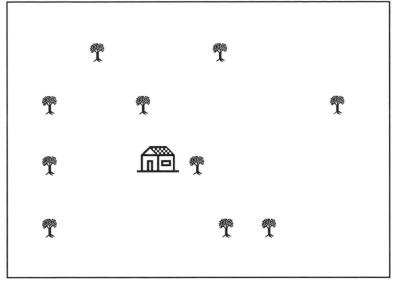

Figure 7.5

Again, trying to solve this puzzle with accurate reckoning leads nowhere. On the other hand, envisioning the critical elements of the puzzle produces the relevant insight:

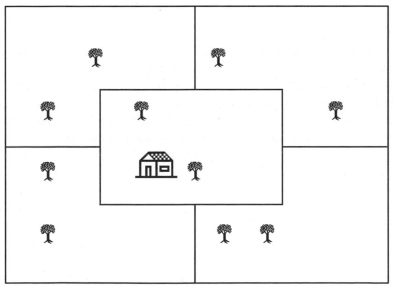

Figure 7.6

In practice, accurate reckoning and insight thinking are both necessary cognitive strategies, and must be used together in the solution of all kinds of problems. In his "Murders in the Rue Morgue" Edgar Allan Poe called this blend of thinking modes the human being's "bi-part soul," which, he suggested, produces in all of us the mind of a "poet-mathematician." The bi-part soul is the source of all great discoveries in science and mathematics.

A classic example of bi-part thinking is Euclid's proof that the prime numbers are infinite. The integers are divided into those that can be decomposed into factors—*composite numbers*—and those that cannot. The numbers 12, 42, and 169, for instance, are all composite, because they are the products of smaller numbers, called factors: e.g., $12 = 2 \times 2 \times 3$, $42 = 7 \times 2 \times 3$, $169 = 13 \times 13$. The *prime* numbers are those that cannot be decomposed in this way. The factors in the examples above—2, 3, 7, and 13—are prime. (The number 1 is a special case, neither prime nor composite.) The first ten primes are 2, 3, 5, 7, 9, 11, 13, 17, 19, 23. Now, even a cursory examination of the number line—{0, 1, 2, 3, 4, 5, 6, 7, 8, 9, 10, . . .}—reveals that there are fewer and fewer primes as the numbers increase. Thus, it appears logical to conclude that the primes must come to an end at some point. Common sense would also have it that if a number is big enough, it must be the product of other, smaller numbers. But, with a blend of insight thinking and accurate reckoning, Euclid proved that this is not so.

He started with the assumption that there may indeed be a finite set of primes, labeling them as follows: $\{p_1, p_2, p_3, \ldots p_n\}$. The symbol p_n stands for the last (largest) prime. Concretely, the set would look like this: $\{2, 3, 5, 7, 9, \ldots p_n\}$. Euclid then obviously had one of the illuminations mentioned above: What kind of number would result from multiplying all the primes in the set: $p_1 \times p_2 \times p_3 \times \ldots \times p_n$? The result would, of course, be a composite number, because it can be factored into smaller prime factors—p_1, p_2, etc. Then Euclid added 1 to this product: $(p_1 \times p_2 \times p_3 \times \ldots \times p_n) + 1$. Now, this number is not decomposable, because when any of the prime factors available to us (p_1, p_2, p_3, $\ldots p_n$) are divided into it, a remainder of 1 will always be left over. So the number $(p_1 \times p_2 \times p_3 \times \ldots p_n) + 1$ is either (1) a prime number that is, obviously, much greater than p_n, or (2) a composite number with a prime factor that, as just argued, cannot be found in the set $\{p_1, p_2, p_3, \ldots p_n\}$ and is thus also greater than p_n. Either way, there must always be a prime number greater than p_n. In a phrase, Euclid showed that the primes never end.

Euclid's proof puts on display the power and mysterious appeal of bi-part thinking. Euclid complemented his *Aha!* insight (adding 1 to the product of the supposed set of primes) with the method known as *reductio ad absurdum*, an indirect method of proof that establishes the truth of something by showing that its contradiction is either false or inconsistent. Incidentally, ever since its publication, Euclid's proof has motivated myriads of mathematicians to come up with a formula for generating the prime numbers—but so far to no avail. One of these was Eratosthenes of Alexandria (chapter 1), whose method came to be known as the "sieve of Eratosthenes." His method proceeds as follows. The first 100 digits are laid out in a square 10×10 grid:

1	2	3	4	5	6	7	8	9	10
11	12	13	14	15	16	17	18	19	20
21	22	23	24	25	26	27	28	29	30
31	32	33	34	35	36	37	38	39	40
41	42	43	44	45	46	47	48	49	50
51	52	53	54	55	56	57	58	59	60
61	62	63	64	65	66	67	68	69	70
71	72	73	74	75	76	77	78	79	80
81	82	83	84	85	86	87	88	89	90
91	92	93	94	95	96	97	98	99	100

Figure 7.7

After the first prime is identified, which is 2, all multiples of 2 (which of course are not prime, since they have 2 as a factor) are crossed out. Similarly, after the second prime number, 3, is identified, all its multiples are crossed out. This process is continued until only the primes among the first 100 numbers are left in the sieve. Of course, the number 1 is also crossed out (since it is neither prime nor composite).

This method generates the first 25 primes: 2, 3, 5, 7, 11, 13, 17, 19, 23, 29, 31, 37, 41, 43, 47, 53, 59, 61, 67, 71, 73, 79, 83, 89, and 97. From 100 to 200, there are 21 primes. From 200 to 300 there are 16. And so it goes: the primes become more sparse the higher up the number line one goes.

The French mathematician Marin Mersenne (1588–1648) also sought a formula that would generate all the primes. Mersenne failed, of course, to come up with one. But the one that he proposed, $2^n - 1$, nevertheless became important to the development of *number theory*—the branch of mathematics studying number properties. The numbers generated by his formula are known, appropriately enough, as *Mersenne numbers:*

$2^2 - 1 = 3$
$2^3 - 1 = 7$
$2^4 - 1 = 15$
$2^5 - 1 = 31$
$2^6 - 1 = 63$
$2^7 - 1 = 127$

etc.

Some of these (e.g., 3, 7, 31, and 127) are indeed prime, and are thus called *Mersenne primes.* Mersenne showed that his formula generates the following large primes:

$2^{13} - 1 = 8,191$
$2^{17} - 1 = 131,071$
$2^{19} - 1 = 524,287$

He further claimed that $2^{31} - 1$, $2^{67} - 1$, and $2^{257} - 1$ were prime. The first one was proven to be so in 1772, but the latter two turned out to be composite.

Many other illustrious mathematicians can be added to the list of those who have attempted, unsuccessfully, to devise a formula for generating all the primes. Since 1952, computers have become the primary tools in the search for primes. In 1978, two high school students in California, Laura Nickel and Curt Landon Noll, found that $2^{21,701} - 1$ was prime, using computer techniques (Peterson and Henderson 2000: 37). It was the twenty-fifth Mersenne prime to be discovered. It has 6,533 digits. In 1996, a loose international Internet alliance of prime number aficionados was founded by computer programmer George Woltman in Florida. Known as GIMPS (the Great Internet Mersenne Prime Search Project), it has brought together more than four thousand devotees in a systematic effort to search for primes. GIMPS has established the largest Mersenne prime known: $2^{3,021,377} - 1$. It is the thirty-seventh Mersenne prime to be discovered and has 909,526 digits.

Prime numbers are the building blocks of the whole architecture of arithmetic. In a letter to Euler in 1742, the mathematician Christian Goldbach (1690–1764) conjectured that every even integer greater than 2 could be written as a sum of two primes (Salem, Testard, and Salem 1992: 110–111):

$$4 = 2 + 2 \qquad 8 = 5 + 3 \qquad 12 = 7 + 5 \qquad 16 = 11 + 5$$
$$6 = 3 + 3 \qquad 10 = 7 + 3 \qquad 14 = 11 + 3 \qquad 18 = 11 + 7$$

etc.

No exception is known to Goldbach's Conjecture, as it has come to be known, but we have no valid proof of it. Goldbach also conjectured that any number greater than 5 could be written as the sum of three primes:

$$6 = 2 + 2 + 2 \qquad 8 = 2 + 3 + 3 \qquad 10 = 2 + 3 + 5$$
$$7 = 2 + 2 + 3 \qquad 9 = 3 + 3 + 3 \qquad 11 = 3 + 3 + 5$$

etc.

In his delightful novel *Uncle Petros and Goldbach's Conjecture* (2000), the Greek writer Apostolous Doxiadis treats Goldbach's Conjecture as one of those revelations provided occasionally by God to mystify human ingenuity, even if it is doubtful that, should an explanation for the conjecture ever be revealed, it would change the world in any way.

But some "revelations" have indeed changed the world. Take, for example, the Euclidean postulate that through a point outside a given line it is possible to draw only one line parallel to the given line—i.e., one that will never meet the given line no matter how far the lines are extended in either direction. On a sheet of paper, this postulate is true— line A and line B will never meet, no matter how far they are extended to the left or to the right:

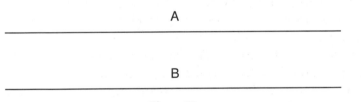

Figure 7.8

In the first part of the nineteenth century the German mathematician Carl Friedrich Gauss, the Russian mathematician Nicolai Ivanovich Lobachevsky (1792–1856), and the Hungarian mathematician János Bolyai (1806–1860) independently had the same capricious hunch. Might there be a "world" where Euclid's postulate is untrue? And, as God revealed, to paraphrase Doxiadis, indeed there is! The area inside a circle is one such realm. There, an infinite number of parallels to line A can be drawn through the point P. The reason for this is, of course, that those lines, being inside the circle, cannot be extended beyond its circumference.

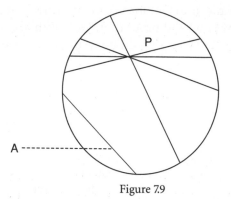

Figure 7.9

Of course, if the lines were to be extended to the world outside the circle, along with A, then all of them would intersect A. Around 1860, the German mathematician Georg Friedrich Bernhard Riemann (1826–1866) had another whimsical hunch. Might there be a world where no lines are parallel? Once again, as God revealed, there is such a world— the surface of a sphere, on which straight lines are defined as great circles, whose planes pass through the center of the sphere. It is, in fact, impossible to draw any pair of nonintersecting great circles on the surface of a sphere. So, if A and B above were transferred to the surface of an ordinary globe (as shown), they would meet at the two poles:

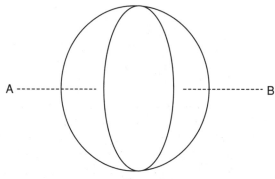

Figure 7.10

Incidentally, this insight is the basis for a classic puzzle that never fails to stump first-time solvers (Kasner and Newman 1940: 145):

> *A group of sportsmen, having pitched camp, set forth to go bear hunting. They walk 15 miles due south, then 15 miles due east, where they sight a bear. They return to camp by traveling 15 miles due north. What was the color of the bear?*

How can the sportsmen travel as stipulated and end up back at the camp? On a two-dimensional surface this is, of course, impossible:

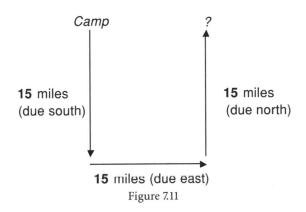

Figure 7.11

But the earth's surface is spherical, not planar. The camp is pitched at the North Pole, and the travel directions described by the puzzle will lead the sportsmen back to the pole, no matter how far east they go. Hence the bear is a polar bear, which is white.

For comparatively small distances, such as those we experience every day, Euclidean and non-Euclidean geometries are essentially equivalent. However, in dealing with astronomical space and such problems of modern physics as relativity and the theory of wave propagation, non-Euclidean geometries give a more precise description of the observed phenomena than does Euclidean geometry.

Puzzles and Serendipity

A second theme that has guided us on our trek through Puzzleland has been the fact that puzzles have often led serendipitously to mathematical discoveries. As we have seen in previous chapters, Euler's Königsberg's Bridges Puzzle led to the establishment of network theory, Alcuin's River-Crossing Puzzle provided the basic insight for developing critical path analysis, and so on. Mathematicians are sometimes so inspired by a problem that, as Dehaene (1997: 151) puts it, "truth descends upon them," leading them to imagine some new hypothesis, conjecture, or theory. The most famous example of this is provided by the French mathematician Pierre de Fermat. While working on puzzles in Diophantus's *Arithmetica,* which had been translated into French by the puzzlist Bachet, Fermat became keenly interested in Pythagorean triples—sets of three numbers, a, b, and c, for which the equation $a^2 + b^2 = c^2$ is true. This equation reflects, of course, the fact that the square on the hypotenuse of a right triangle (c) is equal to the sum of the squares on the other two sides (a and b):

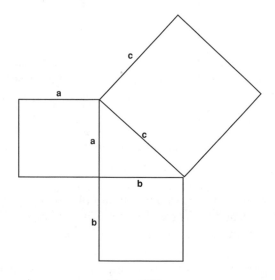

Figure 7.12

For the sake of historical accuracy, it should be mentioned that this relation was known far and wide before it was proved. Clay tablets recovered by archaeologists, dating back to nearly 2000 B.C., reveal that the ancient Babylonians knotted ropes to make 3–4–5 right triangles—because $3^2 + 4^2 = 5^2$ (Neugebauer, Sachs, and Götze 1945). They clearly had a practical knowledge of what later came to be known as the Pythagorean Theorem, and they were also familiar with many Pythagorean triples— e.g., 3, 4, 5 ($3^2 + 4^2 = 5^2$); 6, 8, 10 ($6^2 + 8^2 = 10^2$); 5, 12, 13 ($5^2 + 12^2 = 13^2$); 8, 15, 17 ($8^2 + 15^2 = 17^2$); etc. Fascination with such triples was widespread in the ancient world. A clay tablet dating back to 1900 B.C.—

referred to as *Plimpton 322*—contains fifteen numbered lines with two figures in each line that are Pythagorean (Wells 1992: 7). Moreover, proofs of the Pythagorean Theorem may even predate Pythagoras. One of the most elegant can be found in a Chinese manuscript titled *Chou Pei*, which some believe can be dated back to 1200 B.C., although others estimate that it was written around A.D. 100 (Li Yan and Du Shiran 1987). This astonishingly simple proof of what the Chinese called *Gougou* is presented as two self-explanatory diagrams:

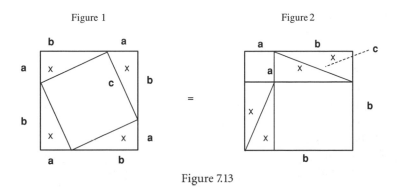

Figure 7.13

The length of the hypotenuse c in triangle abc in Figure 2 is used to draw a square with sides c in Figure 1, around which a rectangular figure is drawn with sides of length $(a + b)$, as shown. The corresponding four triangles in each figure (marked x) are equal. Subtracting the four equal triangles from both figures leaves the areas $a^2 + b^2 = c^2$.

In ancient Greece, the Pythagorean numbers captivated the imagination of many more mathematicians than just the Pythagoreans. Diophantus included a long, insightful discussion of them in his *Arithmetica*. He also devised puzzles whose solution required the Pythagorean Theorem. It was that discussion and those puzzles that stimulated Fermat's imagination. In the margin of his copy of Diophantus's book, he wrote the following enigmatic words (cited by Pappas 1991: 150):

> To divide a cube into two cubes, a fourth power, or in general any power whatever above the second, into two powers of the same denomination, is impossible, and I have assuredly found an admirable proof of this, but the margin is too narrow to hold it.

Fermat claimed that his proof would show, in effect, that only for the value $n = 2$ do solutions of $a^n + b^n = c^n$ exist: namely, $a^2 + b^2 = c^2$. (The trivial case $a = b = c = 1$ is discounted.) For more than 350 years, mathematicians across the world were intrigued by Fermat's claim, trying valiantly to come up with a proof, but always to no avail, although a number of special cases were settled. Gauss proved that $a^3 + b^3 = c^3$ had no positive solutions, and Fermat himself proved the untenability of $a^4 + b^4 = c^4$. The French mathematician Adrien Marie Legendre (1752–1833) gave a proof that $a^5 + b^5 = c^5$ had no solutions. And A. Dirichelt (1805–1859) showed that $a^{14} + b^{14} = c^{14}$ had no solutions. But no general proof, as Fermat envisioned it, was discovered until in June 1993 Andrew Wiles (b. 1953), an English mathematician teaching at Princeton University, declared that he had finally proven what had come to be known as Fermat's Last Theorem. In December of that year, however, some mathematicians found a gap in his proof. In October 1994 Wiles, together with another English mathematician, Richard L. Taylor (b. 1962), filled that gap to virtually everyone's satisfaction. The Wiles-Taylor proof was published in May 1995 in the *Annals of Mathematics*.

But Fermat's Last Theorem still haunts mathematicians, for the simple reason that the Wiles-Taylor proof was certainly not what Fermat envisioned. Their proof required a computer program and depended on mathematical work subsequent to Fermat. In a pure sense, therefore, the Wiles-Taylor proof does not really constitute a historically appropriate resolution to Fermat's Last Theorem. Fermat left behind a true mathematical detective mystery. What possible simple proof could he have been thinking of as he read Diophantus's *Arithmetica*? There are undoubtedly some mathematicians still trying to find Fermat's mysterious proof, if indeed he ever had one in the first place. As Ian Stewart (1987: 48) aptly puts it, "Either Fermat was mistaken, or his idea was different." In my view, mathematicians will probably not rest until this "obscurest of secrets," to paraphrase Ahmes once again, is brought finally to the light of reason.

One line of inquiry that, to the best of my knowledge, has never been pursued in investigating Fermat's Last Theorem is in the domain of equation systems and of their reference to dimensions in terms of exponents. As Diophantus asserted, the Pythagorean relation holds because it links one number (the square of the hypotenuse) to two others (the squares of the other two sides). This relation is reified by a figure in two-dimensional space, namely, a right-angled triangle:

$$a^2 + b^2 = c^2$$

The c^2 in this equation can be called the *Pythagorean term*, for the sake of convenience, and represented with p^n. In the above case $n = 2$, which of course indicates the number of dimensions to which the Pythagorean relation refers. It would therefore seem logical to assume that in three-dimensional space, the corresponding relation would have to take into account this new dimension. This could thus be represented by one more variable in the left side of the equation, corresponding to the number of dimensions, $n = 3$:

$$a^3 + b^3 + c^3 = p^3$$

Following this line of reasoning, in n-dimensional space the Pythagorean relation could be expressed as follows:

$$a^n + b^n + c^n + d^n + \ldots + n^n = p^n$$

This shows that a variable is added to the equation in accordance with the number of dimensions. In a seven-dimensional space, where $n = 7$, there would therefore be seven variables in the left side of the equation:

$$a^7 + b^7 + c^7 + d^7 + e^7 + f^7 + g^7 = p^7$$

It is certainly beyond the scope of the present discussion to go into the merits or demerits of this proposal. Let me suggest that it presents some intriguing possibilities for reconsidering Fermat's Last Theorem. What kind of seven-dimensional figure would constitute the reification of the above formula? Could Fermat's proof have been consistent with this line of inquiry?

The Wiles-Taylor proof is the result of a synthesis of previous work. But, as far as I can tell, it leaves many mathematicians unsatisfied. The question that seems to linger in their minds is this: Did Fermat, through a flash of insight, stumble serendipitously upon some truly mysterious hidden pattern that defies logic with its apparent simplicity?

The word "serendipity" was coined by Horace Walpole in 1754, from the title of the Persian fairy tale "The Three Princes of Serendip," whose heroes make many fortunate discoveries accidentally. Serendipity characterizes the history of mathematics. Consider, for instance, Pascal's Triangle, named after the French mathematician Blaise Pascal, one of the founders of probability theory (chapter 6). Pascal generated each number in his triangle by adding the two numbers just above it:

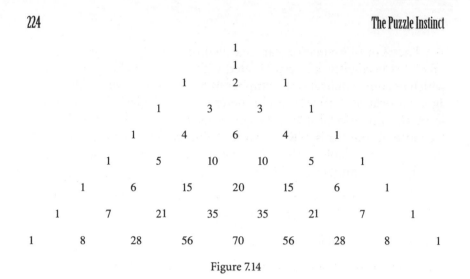

Figure 7.14

For example, the 70 in the bottom row is equal to 35 + 35—the sum of the two numbers that are just above it. Similarly, the 6s in the third row from the bottom are equal to 5 + 1—the sum of the two numbers just above them. For the sake of historical accuracy, it should be mentioned that Pascal's triangle was known as early as 1303, since it is found in the *Precious Mirror of the Four Elements,* published in that year by the Chinese mathematician Chu Shih-chieh (Gullberg 1997: 141).

The amazing thing about this triangle—which, of course, can be enlarged indefinitely—is that it appears serendipitously as a hidden pattern in other areas of mathematics. It surfaces, for instance, in the expansion of the expression $(a + b)^n$, which is described by the *binomial theorem,* discovered in the West by the great scientist Isaac Newton when he was a student in his early twenties—although the theorem was already common knowledge in China in the fourteenth century (Sardar, Ravetz, and Van Loon 1999: 67). In the expansion of $(a + b)^4$, which is $a^4 + 4a^3b + 6a^2b^2 + 4ab^3 + b^4$, the coefficients of the terms (1, 4, 6, 4, 1) coincide with the fourth row of numbers in Pascal's triangle. Similarly, the coefficients in the expansion of $(a + b)^5$ coincide with the fifth row of numbers in the triangle; those in the expansion of $(a + b)^6$ with the sixth row; and so on. Generally speaking, the coefficients of the terms in the expansion of $(a + b)^n$ coincide with the nth row of numbers in Pascal's triangle:

$(a + b)^0 = 1$
$(a + b)^1 = a + b$
$(a + b)^2 = a^2 + 2ab + b^2$

$(a+b)^3 = a^3 + 3a^2b + 3ab^2 + b^3$
$(a+b)^4 = a^4 + 4a^3b + 6a^2b^2 + 4ab^3 + b^4$
$(a+b)^5 = a^5 + 5a^4b + 10a^3b^2 + 10a^2b^3 + 5ab^4 + b^5$
$(a+b)^6 = a^6 + 6a^5b + 15a^4b^2 + 20a^3b^3 + 15a^2b^4 + 6ab^5 + b^6$
$(a+b)^7 = a^7 + 7a^6b + 21a^5b^2 + 35a^4b^3 + 35a^3b^4 + 21a^2b^5 + 7ab^6 + b^7$
$(a+b)^8 = a^8 + 8a^7b + 28a^6b^2 + 56a^5b^3 + 70a^4b^4 + 56a^3b^5 + 28a^2b^6 + 8ab^7 + b^8$
etc.

Pascal's numbers also crop up in the calculation of probabilities. If a coin is to be tossed eight times, the possible outcomes are as follows (H = heads, T = tails). One outcome consists only of heads:

H H H H H H H H

For seven heads and one tail, there are 8 favorable outcomes:

H H H H H H H T
H H H H H H T H
H H H H H T H H
H H H H T H H H
H H H T H H H H
H H T H H H H H
H T H H H H H H
T H H H H H H H

For six heads and two tails, there are 28 favorable outcomes (as readers may wish to verify for themselves); for five heads and three tails, there are 56 favorable outcomes; and so on. These outcomes thus coincide with the numbers in the eighth row of Pascal's triangle (1, 8, 28, 56, 70, 56, 28, 8, 1): i.e., there is 1 outcome of no tails; 8 outcomes of one tail; 28 outcomes of two tails; and so on. Altogether, the number of possible outcomes is:

$1 + 8 + 28 + 56 + 70 + 56 + 28 + 8 + 1 = 256$

In general, the numbers in the nth row of Pascal's triangle correspond to the probability of each possible outcome of n tosses of a coin. So the probability of getting all heads in eight tosses is 1/256; the probability of seven heads and one tail is 8/256 = 1/32; the probability of six heads and two tails is 28/256 = 7/64; and so on.

But perhaps the most remarkable serendipity of all is that the diagonal sums of the numbers in Pascal's triangle correspond to the numbers in the Fibonacci sequence (1, 2, 3, 5, 8, 13, 21, 34, . . .) (see also Gardiner 1987: 70; Vernadore 1991):

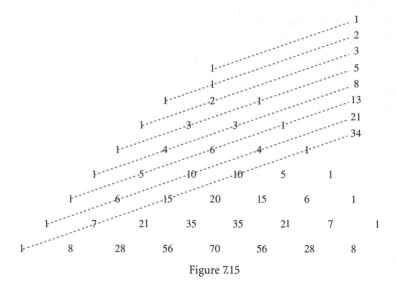

Figure 7.15

It staggers the mind to contemplate such remarkable coincidences. Why do Pascal's and Fibonacci's numbers correspond in this way? Why does Pascal's triangle forecast chance outcomes? The mind boggles to find answers to such questions. They remain mysteries to this day.

The topic of serendipity is taken up by writer and semiotician Umberto Eco in his engrossing book *Serendipities: Language and Lunacy* (1998). Eco argues that many of the great scientific discoveries of history were not made through the application of some systematic empirical method. More often than not, they were serendipitous. The discovery of America by Christopher Columbus (1451–1506) is a case in point. Columbus's assumption that the world was smaller than it actually turned out to be led him to seek a quick route to the East via the West. Thus, by pure serendipity, Columbus found America! The ancient concept of Fortune, Eco observes, served to explain serendipitous events. Fortune was thought of as a personified force that favorably or unfavorably governed human events and discoveries. This is why the ancient scientists consulted oracles and paid homage to Fortune for their inventions and discoveries.

Puzzles and the Aesthetics of Mind

A third theme that has guided us on our foray through Puzzleland is that puzzles are pleasurable in themselves. The suspense that accompanies an attempt to find a solution to a challenging puzzle, or the anxiety that develops from not finding one right away, is a significant part of what makes the puzzle so fascinating and engaging. Like detective stories, puzzles are solved by a mixture of imagination and logic or, as Peirce would have it, *logica utens* and *docens*. And, like such stories, puzzles often constitute a form of escapism.

The peculiar kind of pleasure that puzzles produce can be called an *aesthetics of mind*. The word "aesthetics" requires some commentary. It means, literally, "perceiving with all the senses." It is used in art criticism to refer to the sense of beauty or the emotional feeling of meaningfulness that ensues from the experience of an artistic work. Poetry and music, for instance, evoke a cathartic response that imparts a sense of meaningfulness to existence. This can be called, more specifically, an *aesthetics of emotion*. One does not have to know musical theory or technique in order to recognize the beauty and emotional force of a piano concerto by Mozart. Mozart's art triggers our senses and emotions directly. Puzzles cannot do that. They can never be characterized as sad or happy; they can only be called ingenious or clever. But indulging in ingenuity and cleverness can, as we have seen in our journey through Puzzleland, produce a form of pleasure nonetheless. As the British puzzlist Hubert Phillips (1937: vii) put it, solving puzzles provides an intellectual "kick," which results from discovering the patterns, traps, or tricks they conceal. This is true for people of all ages, as borne out by the fact that puzzles of all kinds appeal to children and adults alike.

Needless to say, some puzzles are more intellectually pleasurable than others are. The *aesthetic index* of a puzzle, as it may be called, seems to be inversely proportional to the complexity of its solution or to the obviousness of the pattern, trap, or trick it hides. Simply put, the longer and more complicated the answer to a puzzle, or the more obvious it is, the less appealing the puzzle seems to be. Puzzles with simple yet elegant solutions, or puzzles that hide a nonobvious principle, have a higher aesthetic index. As John Allen Paulos argues in his fascinating books (1980, 1985, 1991: 113), puzzles are forms of intellectual play, based on the same mental plan in which humor is rooted. In both cases, getting to the "punch line" is the source of the pleasure. The less obvious the punch line, the funnier the joke; the less obvious the answer, the more pleasurable the puzzle.

The aesthetic index is also very high when solving a puzzle or contemplating some mathematical proof or demonstration produces a paradoxical result. An example of this is Cantor's demonstration that, when dealing with infinite sets, the whole turns out to be equal to parts of itself (chapter 4)—a paradox, incidentally, that crossed the mind of Galileo in his 1632 *Dialogue Concerning the Two Chief World Systems* and that was suspected by such notable mathematicians as Gottfried Leibniz, Bernhard Bolzano (1781–1848), Karl Weierstrass (1815–1907), and J. W. R. Dedekind (1831–1916). Galileo pointed out, in fact, that there are as many squares as there are natural numbers, because they are just as numerous as their roots. In view of the fact that there are natural numbers that are not squares—2, 3, 5, 6, 7, 8, 10, 11, 12, etc.—Galileo's statement appears to be a paradox.

Following on Galileo's coattails, Cantor examined this relation in depth. He started by defining the two sets:

Set A = all values of n, where n is an integer:

$\{1, 2, 3, 4, 5, 6, 7, 8, 9, \ldots\}$

Set B = all values of n^2, where n is an integer:

$\{1, 4, 9, 16, 25, 36, 49, 64, 81, \ldots\}$

At first glance, there seem, logically, to be more elements in set A than in set B. After all, we can put them into a one-to-one correspondence as follows, with A on top and B on the bottom:

1	2	3	4	5	6	7	8	9	10	11	12	and so on
↕	↕	↕	↕	↕	↕	↕	↕	↕	↕	↕	↕	
1	-	-	4	-	-	-	-	9	-	-	-	and so on

Figure 7.16

As expected, there are many more "blanks" in set B, because it is a subset of A. It thus seems logical to conclude that set A—the whole—has many more members than set B—a part of set A. But it does not. Following Galileo's suggestion and subsequent work by the German mathematician Dedekind in 1872, Cantor showed that the two sets have the same number of elements. He did this simply by eliminating the blanks:

1	2	3	4	5	6	7	8	9	10	11	12	and so on
↕	↕	↕	↕	↕	↕	↕	↕	↕	↕	↕	↕	
1	4	9	16	25	36	49	64	81	100	121	144	and so on

Figure 7.17

This system of matching shows that, no matter how far the number line is extended, every member of the top set will always be matched with one and only one member of the bottom set, and vice versa. So, as it turns out, the two sets have the same number of elements, a result that seems to defy common sense.

Such paradoxes surprise and delight us at the same time, generating a high aesthetic index, because they produce unexpected and even counterintuitive results. As Kasner and Newman (1940: 43) aptly observe, Zeno would have been proud of Cantor, and would not have challenged Cantor's demonstration "in spite of his skepticism about the obvious." Remarkably, Cantor's proof applies as well to other infinite subsets of the integers. For instance, there are as many even numbers and as many odd numbers as there are numbers (as we discussed briefly in chapter 4):

integers	1	2	3	4	5	6	7	8	9	10	11	12
	↕	↕	↕	↕	↕	↕	↕	↕	↕	↕	↕	↕
even integers	2	4	6	8	10	12	14	16	18	20	22	24

Figure 7.18

integers	1	2	3	4	5	6	7	8	9	10	11	12
	↕	↕	↕	↕	↕	↕	↕	↕	↕	↕	↕	↕
odd integers	1	3	5	7	9	11	13	15	17	19	21	23

Figure 7.19

Are any other infinite subsets of numbers equal in number to the integers? Consider the set of rational numbers. These are numbers that can be written in the form p/q, where p and q are integers. Thus, for instance, ⅔, –⅝, 5, and 4/7 are rational numbers. Clearly, the integers are a subset of the rationals, because every integer n can be written in the form $n/1$. Terminating decimal numbers are also rational numbers, because a number such as 3.579 can be written in p/q form as 3579/1000. Finally, all repeating decimal numbers are also rational, although the proof of this is beyond the scope of this book. For example, 0.3333333 . . . can be written as ⅓. An important property of the rational numbers, Cantor noticed, is that they are *dense*: i.e., between any two rational numbers there are infinitely many other rational numbers. For example, between 0 and 1 there exist an infinite number of fractions.

Amazingly, Cantor demonstrated that the number of rationals is also equal to the number of integers. His method of proof is, again, unexpectedly elegant and simple. First, he arranged the set of all rational numbers as shown in the figure below:

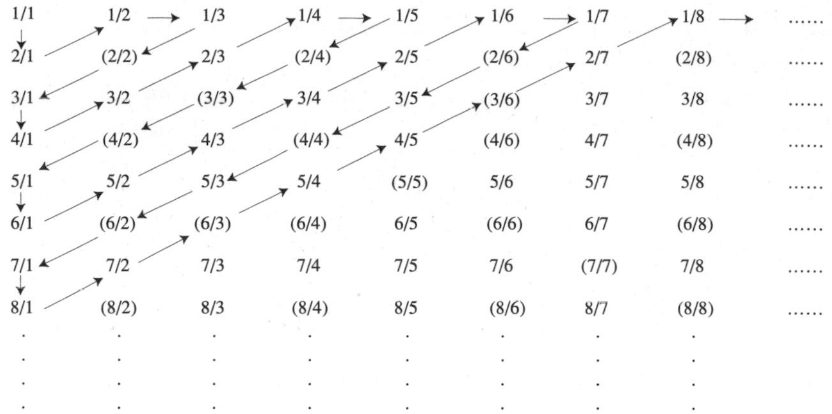

Figure 7.20

In each row the successive denominators (q) are the integers $\{1, 2, 3, 4, 5, 6, \ldots\}$, while all the numerators (p) in the first row are 1, those in the second row are all 2, those in the third row are all 3, and so on. Cantor enclosed every fraction in which the numerator and the denominator have a common factor in parentheses. If these fractions are deleted, then every rational number appears once and only once in the array, which is called *Cantor's sieve*, since it is evocative of Eratosthenes' sieve, discussed above. Now, Cantor set up a one-to-one correspondence between the integers and the numbers in his sieve as follows. First he let 1 correspond to ⅟₁ at the top left-hand corner of the sieve; then he let 2 correspond to the number below (²⁄₁); following the arrow path, he let 3 correspond to ½; after that, following the arrow path, he let 4 correspond to ⅓; and so on, ad infinitum. Eliminating the numbers in parentheses, the following one-to-one correspondence is thus generated by his sieve:

integers	1	2	3	4	5	6	7	8	9	10	11	12	13	...
	↓	↓	↓	↓	↓	↓	↓	↓	↓	↓	↓	↓	↓	
sieve numbers	1/1	2/1	1/2	1/3	3/1	4/1	3/2	2/3	1/4	1/5	5/1	6/1	5/2	...

Figure 7.21

This shows plainly that to every rational number there corresponds one and only one integer, and to every integer there corresponds one and only one rational number. This result truly confounds and pleases us at the same time. Our aesthetic response to Cantor's proof lies not only in having encountered a paradox, but also in the elegant and simple way in which he constructed his demonstration. As Kasner and Newman (1940: 56) aptly put it, once the simplicity inherent in the principles of Cantor's demonstrations and the overall theory are understood, they "cease to sound like the extravagances of a mathematical madman."

Yet we are completely ignorant of the underlying reasons for such patterns, and we shall perhaps always remain ignorant of them. This is why so many puzzles trace their origins to portentous or knowledge-testing events, and why some, like anagrams, labyrinths, and magic squares, have been thought to constitute metaphors for life and destiny from time immemorial. Buried deeply within the imagination is the belief that if we were to solve all the puzzles of the world we could bring about change in that world, because we would have discovered the mystical structures that it conceals.

Ahmes' Challenge

The theme of mystery, which has guided a large portion of our journey through Puzzleland, is a key element in the Rhind Papyrus. As Ahmes explicitly states, puzzles provide a key for gaining "entrance into the knowledge of all existing things and all obscure secrets." This association between puzzles and mystery has become largely unconscious in modern people, but it is there nonetheless. This is both Ahmes' legacy and his challenge. We continue to solve puzzles, as if we were impelled by some ancient force or magical spell to do so. Yet, as we have seen, puzzles are not merely the product of superstition. More often than not they have led to important discoveries in mathematics and science, as well as to serious reconsiderations of philosophical methods.

The universal fascination with puzzles raises an important question. Do they serve some hidden survival function? In many parts of academia today, a debate is going on that is relevant to this very question. Academicians like Desmond Morris (e.g., 1969), Richard Dawkins (e.g., 1976, 1987, 1995, 1998), Robert Wright (1994), and Stephen Pinker (1997) argue, with great rhetorical deftness and aplomb, that human symbols and texts enhance survivability and promote progress by replacing the functions of genes with *memes*—a word coined by Dawkins. Memes are replicating patterns of information (ideas, laws, clothing fashions, artworks, puzzles, etc.) and of behavior (marriage rites, love rituals, religious ceremonies, etc.) that people inherit directly from their cultural environments. Like genes, memes involve no intentionality on the part of the receiving human organism. Being part of culture, the human organism takes them in unreflectively from birth, passing them on just as unreflectively to subsequent generations. The *memetic code* has thus replaced the *genetic code* in directing human evolution. This clever proposal poses an obvious challenge to virtually everything that has been written on human nature in traditional philosophy, theology, and science. If scholars like Dawkins are correct, then the search for meaning to existence beyond physical survival is essentially over. Any attempt to seek spiritual meaning in life would be explained as one of the intellectual effects of culturally inherited memes such as the soul, God, and immortality. Memes have developed simply to help human beings cope with their particular form of consciousness, thus enhancing their collective ability to survive as a species (Segerstråle 2000).

The key figure behind this whole intellectual movement is the North American biologist E. O. Wilson (b. 1929), known for his work tracing the effects of natural selection on biological communities, especially on populations of insects, and for extending the idea of natural selection to human cultures. Since the mid-1950s, Wilson has maintained that social behaviors in humans are genetically based and that evolutionary processes favor those behaviors that enhance reproductive success and survival. Thus, characteristics such as heroism, altruism, aggressiveness, and male dominance should be understood as evolutionary outcomes, not in terms of social or psychic processes. Moreover, Wilson sees the creative capacities undergirding language, art, scientific thinking, myth, etc. as originating in genetic responses that help the human organism survive and continue the species. As he has stated rather bluntly, "no matter how far culture may take us, the genes have culture on a leash" (Wilson and Harris 1981: 464).

Obviously captivated by such rhetoric, popular magazines and TV science programs have conveyed this new paradigm to large audiences with simple examples. But the study of puzzles throws an intellectual monkey wrench, so to speak, into this whole train of thought. What, one might ask, do puzzles have to do with survival or species continuity? What kinds of animal mechanisms are they remnants of? Like all theories, memetic theory is itself the product of a particular worldview. To paraphrase the French philosopher Michel Foucault (1926–1984), human beings seek to understand and define their reality by ascribing it to either Nature, human effort, or God. As others have done in the past, the memetic theorists of today have placed most of their bets on Nature.

Meme theory is, in effect, a contemporary descendant of a philosophical legacy that goes under the rubric of *physicalism*. Although it has ancient roots, this doctrine gained widespread momentum in Western society after the establishment of Darwinian evolutionary biology in the nineteenth century. Conscious social behaviors are, of course, partially based in biology; but they are not totally so (Searle 1999). Genetic factors alone do not completely define human beings. They tell us nothing about why humans create their meaningful experiences and pose the questions they do about life.

Puzzles are "miniature blueprints" of how pattern is wired into the brain and of how we search for answers to existential questions. As philosopher Johan Huizinga (1924: 202) has eloquently put it, these questions have led to the foundation of religious systems: "In God, nothing is empty of sense . . . so, the conviction of a transcendental meaning in all things seeks to formulate itself." As mentioned in the opening chapter, these are large-scale questions that crystallize constantly in our consciousness. Puzzles, on the other hand, are small-scale experiences of these questions.

A search for the meaning of puzzles in human life is tantamount to a search for a meaning to life itself. The great writer T. S. Eliot (1888–1965) argued that true knowledge starts with comprehending the forces that have made us what we are. The puzzle instinct is one of those forces. There is no culture without puzzles; and there is no human being alive who does not understand what a puzzle is. As Henry Dudeney (1958a: 12) aptly put it:

> The curious propensity for propounding puzzles is not peculiar to any race or any period of history. It is simply innate . . . though it is always showing itself in different forms; whether the individual be a Sphinx of Egypt, a Samson of Hebrew lore, an Indian fakir, a Chinese philosopher, a mahatma of Tibet, or a European mathematician makes little difference.

Lewis Carroll, perhaps the greatest puzzlist of all time, expanded both *Alice's Adventures in Wonderland* and its sequel *Through the Looking-Glass* from tales he told to children. Many of his characters— Humpty Dumpty, the Mad Hatter, the March Hare, the White Rabbit, the Red and White Queens—have become familiar figures in literature and conversation. Although numerous satirical and symbolic meanings have been read into Carroll's books, Carroll himself claimed that he meant only nonsense. But the type of nonsense to which Carroll alludes is the same one that motivated Ahmes, Pythagoras, Archimedes, Fibonacci, Cardano, and Euler, to mention but a few. The looking-glass (a mirror) through which Alice must pass in order to get a look at the world on the other side is a marvelous metaphor for the concept of infinity. By holding a mirror to reflect another mirror, an observer will, as if by magic, see a mirror inside a mirror, inside a mirror, and so on without end:

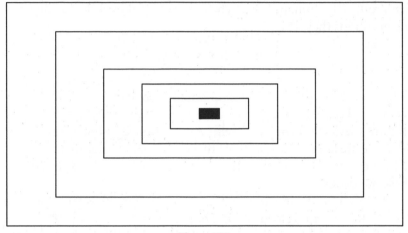

Figure 7.22

By seeing the world through the mirror, Alice derives a different sense of the magnitude of things, both large and small, and thus experiences the same kind of mysticism and awe that the ancient puzzle-makers must have felt. With its universal message and appeal, Carroll's novel inspired many authors, including G. K. Chesterton (1874–1936) and science fiction writer Poul Anderson (b. 1926), whose *The Immortal Game* revolves around a chess game shaped like a looking-glass, in obvious allusion to both *Through the Looking-Glass* and Bergman's *The Seventh Seal* (chapter 6).

There is no simple answer to *Why puzzles?* And this is perhaps why, in a sense, our journey through Puzzleland has led nowhere. In the realm of the imagination there are no linear paths or finite maps that lead to definitive answers. In *Alice's Adventures in Wonderland*, Alice asks the Cheshire cat, "Would you tell me, please, which way I ought to go from here?" The cat's answer is simple, yet revealingly insightful: "That depends a good deal on where you want to get to," to which Alice responds with "I don't much care where." The shrewd cat's rejoinder to Alice's response applies to our own wanderings through Puzzleland: "Then it doesn't matter which way you go."

All that can be said is that puzzles are not just curious figments of mind, but elusive bits of evidence of a theory of the world that is lurking around somewhere, but that seems to evade articulation. I have tried as best I could in this book to present a few of those bits of evidence. Hopefully, I have shed some light on how the puzzle instinct, in its own miniature way, has guided us, and continues to guide us, in our search for an answer, whether real or imagined, to the most vexing puzzle of all—the meaning of life.

SOLUTIONS

Chapter 1

1.1

Using modern algebraic techniques, Metrodorus's puzzle can be solved easily as follows. We start by letting x stand for the amount given to the legitimate son. From this, it follows that $(1000 - x)$ will represent what was given to the illegitimate one, since this expression simply tells us what is left over from 1000 after x staters have been doled out. Now, the puzzle states that the fifth part of the legitimate son's share x, which is expressed as $\frac{1}{5}x$, will exceed the fourth part of the illegitimate son's share $(1000 - x)$, which is expressed as $\frac{1}{4}(1000 - x)$, by 10 staters. In algebraic terms this translates into the following equation:

$$\tfrac{1}{5}x - \tfrac{1}{4}(1000 - x) = 10$$

The solution is $x = 577\frac{7}{9}$ staters. This is what the legitimate son received; the other son thus got $1000 - 577\frac{7}{9}$, or $422\frac{2}{9}$ staters.

1.2

Letting m, w, and c stand for the number of men, women, and children respectively, the statement of the puzzle can be converted into the following two equations:

(1) $m + w + c = 100$

which states in algebraic terms that there are 100 people in all, and

(2) $3m + 2w + \frac{1}{2}c = 100$

which represents the ways in which the 100 bushels are distributed, i.e., m men will receive $3m$ bushels in all, w women $2w$ bushels in all, and c children $\frac{1}{2}c$ bushels in all, for a total of 100 bushels.

Only positive integral values of m, w, and c are permissible, since fractional or negative integral values would have no real-life meaning—people cannot be split into fractions, nor can negative numbers of them exist. Thus c, which stands for the number of children, must be divisible by 2, otherwise $\frac{1}{2}c$ in the second equation would not yield an integer. Algebraically, this can be expressed by replacing c with $2n$, the general form of an even integer (a form that reflects the fact that any number n when multiplied by 2 will always yield an even number). Substituting $c = 2n$ into the equations above produces the following two new ones:

(3) $m + w + 2n = 100$
(4) $3m + 2w + n = 100$

Now, we can multiply equation (4) by 2, yielding the following equivalent equation:

(5) $6m + 4w + 2n = 200$

We can now subtract equation (3) from (5), an operation that reduces the problem to a single equation:

$$6m + 4w + 2n = 200$$
$$- (m + w + 2n = 100)$$
(6) $5m + 3w = 100$

From this, it can be seen that

(7) $w = (100 - 5m)/3$

We note that m, which stands for the number of men, must be less than 99, because, if it were assigned a value of 99 or 100, the total number of people (including women and children) would be greater than 100. So m must be a positive integral value less than 99 which, when substituted into equation (7), $w = (100 - 5m)/3$, will produce a positive integral value for w. If we assign the value 1 to m, a fractional value for w will result (as readers can confirm for themselves). If we let $m = 2$, however, then the value of w turns out to be 30. This is a definite possibility to consider further. So we can go back to one of the two equations above, say (3) $m + w + 2n = 100$, and substitute $m = 2$ and $w = 30$ into it. From this, the value of n turns out to be 34. Now, since $2n = c$, it is obvious that $c = 68$. We now have the solution to the puzzle, since 2 men, 30 women, and 68 children add up to 100 people in all. To check that our solution is correct, we give each of the 2 men 3 bushels, each of the 30 women 2 bushels, and each of the 68 children a ½ bushel. This results in a total of 100 bushels:

2 *men would receive* 2 × 3 *bushels*	=	6 *bushels*
30 *women would receive* 30 × 2 *bushels*	=	60 *bushels*
68 *children would receive* 68 × ½ *bushel*	=	34 *bushels*
	total:	100 *bushels*

1.3

Following is one possible solution to Kirkman's problem, showing how the numerals from 0 to 14 (each one representing a specific girl) can be arranged in five rows of three each within seven sets (each set corresponding to a day of the week) so that no two numerals appear in the same row more than once (Ball 1972: 287) [see figure S.1].

Monday			Tuesday			Wednesday			Thursday		
0	5	10	0	1	4	1	2	5	4	5	8
1	6	11	2	3	6	3	4	7	6	7	10
2	7	12	7	8	11	8	9	12	11	12	0
3	8	13	9	10	13	10	11	14	13	14	2
4	9	14	12	14	5	13	0	6	1	3	9

Friday			Saturday			Sunday		
4	6	12	10	12	3	2	4	10
5	7	13	11	13	4	3	5	11
8	10	1	14	1	7	6	8	14
9	11	2	0	2	8	7	9	0
14	0	3	5	6	9	12	13	1

Figure S.1

1.4

When it is precisely 8:00, the hour hand is at the 8:00 division point and the minute hand at the 12:00 point. A division point is equivalent to a minute on a clock. We start by letting x represent the number of divisions after 8:00 required by the minute hand to overtake the hour hand. The hour hand will be at the x point when the minute hand overtakes it.

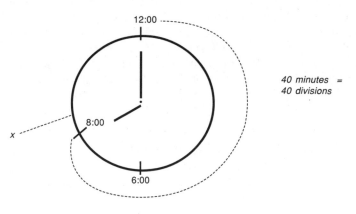

Figure S.2

At 8:00 there are 40 divisions separating the minute hand from the hour hand. To overtake the hour hand, the minute hand will have to cover those 40 divisions plus the extra x divisions after 8:00 that the hour hand has traversed in the meantime. Altogether, it will have to traverse a distance of $(40 + x)$ divisions (= minutes) to overtake the hour hand. The hour hand, of course, will traverse a distance equal to just those x divisions (= minutes). Now, applying the formula *rate = distance/time* (R = D/T) we can represent the rates of the hour and the minute hands as follows:

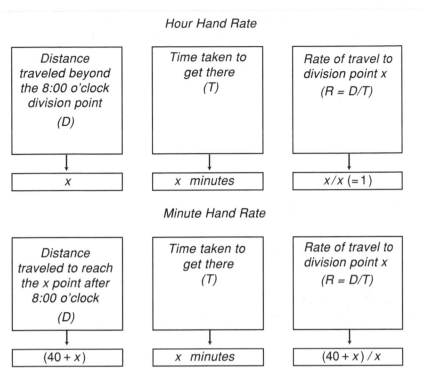

Figure S.3

To connect the two rates, it is necessary to use our practical experience with clocks. We know that the minute hand must cover 60 divisions on the dial in the course of one hour, while the hour hand must cover 5 such divisions. Consequently, the hour hand moves at ⁵⁄₆₀, or ¹⁄₁₂, the rate of the minute hand. This allows us to set up the appropriate equation, as shown below:

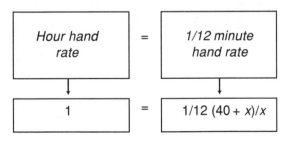

i.e.: 1 = 1/12 (40 + x)/x

Figure S.4

Solving for x in this equation, we get $x = 3\frac{7}{11}$. This means that it will take the minute hand 40 minutes from the 12:00 point to the 8:00 point, plus the $3\frac{7}{11}$ divisions traversed by the hour hand, to overtake the hour hand. In total, therefore, it will take the minute hand $43\frac{7}{11}$ minutes to overtake the hour hand.

Chapter 2

2.1

You can go by steamer to Canada
(name: *Ada*)

This is the white camel I admire most
(name: *Amelia*)

He promises us a nice bottle of wine
(name: *Susan*)

He saw them made at the palace
(name: *Emma*)

2.2

The color names—"yellow," "green," "blue," "white," and "red"—are shown below:

A	B	Y	E	L	L	O	W	N	M
H	J	K	L	M	N	O	P	N	B
V	M	N	N	E	E	R	G	G	L
B	N	M	N	K	O	P	L	K	U
S	W	E	I	B	N	M	K	Y	E
R	T	K	P	W	T	Y	U	M	N
Y	O	P	J	N	H	M	L	M	N
A	D	E	R	Q	U	I	I	E	O
N	M	E	L	P	O	M	T	B	V
X	X	X	R	A	S	D	C	E	X

Figure S.5

2.3

The sentence starts with a one-letter word. Since *A* appears in the plaintext (that is, no cipher symbol is substituted for it here), the only possibility for the initial symbol # is *I:*

*I A$ I*ALIAN*

The solution is now self-evident: *I am Italian.*

2.4

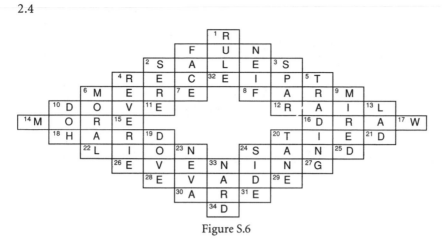

Figure S.6

2.5

Iron
↓
Icon (new letter)
↓
Coin (rearrangement)
↓
Corn (new letter)
↓
Cord (new letter)
↓
Lord (new letter)
↓
Load (new letter)
↓
Lead (new letter)

2.6

M	U	S	I	C				
O	P	E	R	A				
Z	A	I	D	E				
A	U	S	T	R	I	A		
R	E	Q	U	I	E	M		
T	H	I	R	T	Y	–	F	I	V	E

Figure S.7

Chapter 3

3.1

The differences are shown below in Picture *B:*

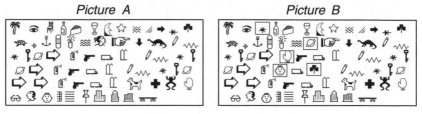

Figure S.8

3.2

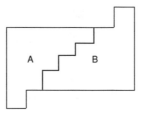

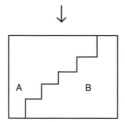

Figure S.9

3.3

The following typical wrong solution, which utilizes "linear logic," as Gardner calls it, shows an obtuse triangle divided into smaller ones. The first three triangles—1, 2, and 3—are acute, but the fourth one turns out to be obtuse. So nothing has been achieved by means of the three cuts.

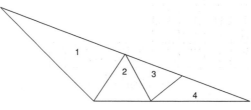

Figure S.10

With a little imagination, however, various solutions, such as the following one with seven acute triangles, are envisionable.

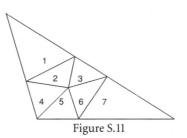

Figure S.11

Chapter 4

4.1

The students involved in the two sports can be represented with two intersecting circles as follows:

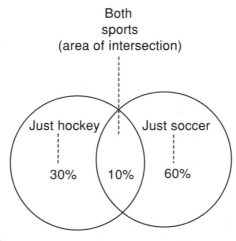

Figure S.12

Jill stated that 70% of the students played soccer, and that 10% played both hockey and soccer. So 60% played only soccer. The intersection of the circles represents the students who played both sports, so it is labeled 10%; we can now label the remainder of the right-hand circle 60%. Since 100% of the grade 10 students played one sport or the other, this means that the remaining 30% played only hockey (100% − 70%), as shown in the remainder of the left-hand circle. In all, 40% of the students played hockey, and Jill stated that there were, in fact, 180 students playing hockey. This number, thus, equals 40% of the total number of students. Consequently, there are 450 students in the grade 10 class, since 40% of 450 = 180.

4.2

We start out on our third logical journey by putting a truth value of F opposite
Jack's first statement, and a T opposite both his second and third statements:

	Statement	Truth Value
Bud	1. I didn't kill Mack.	1. T
	2. Jack is not my friend.	2.
	3. I knew Mack.	3.
Jack	1. I didn't kill Mack.	1. F
	2. Bud and Tug are friends of mine.	2. T
	3. Bud didn't kill Mack.	3. T
Tug	1. I didn't kill Mack.	1. T
	2. Bud lied when he said that Jack was not his friend.	2.
	3. I don't know who killed Mack.	3.

Figure S.13

Bud's second statement can now be seen to be false, because Jack's second statement is true. We can now complete Bud's section of the chart as follows:

	Statement	Truth Value
Bud	1. I didn't kill Mack.	1. T
	2. Jack is not my friend.	2. F
	3. I knew Mack.	3. T
Jack	1. I didn't kill Mack.	1. F
	2. Bud and Tug are friends of mine.	2. T
	3. Bud didn't kill Mack.	3. T
Tug	1. I didn't kill Mack.	1. T
	2. Bud lied when he said that Jack was not his friend.	2.
	3. I don't know who killed Mack.	3.

Figure S.14

Consider again Bud's second statement—*Jack is not my friend*—which we
have established as false. We can now conclude that Tug's second statement—*Bud
lied when he said that Jack was not his friend*—is true, because the chart shows
that Bud did indeed lie when he said that Jack was not his friend. We can safely
put a T opposite Tug's second statement and an F opposite his third statement:

	Statement	Truth Value
Bud	1. I didn't kill Mack.	1. T
	2. Jack is not my friend.	2. F
	3. I knew Mack.	3. T
Jack	1. I didn't kill Mack.	1. F
	2. Bud and Tug are friends of mine.	2. T
	3. Bud didn't kill Mack.	3. T
Tug	1. I didn't kill Mack.	1. T
	2. Bud lied when he said that Jack was not his friend.	2. T
	3. I don't know who killed Mack.	3. F

Figure S.15

So, contrary to what he says—*I don't know who killed Mack*—Tug does indeed know who killed Mack. This contradicts nothing we have deduced so far. In summary, since this third arrangement of Fs and Ts leads to no logical inconsistencies, we can be sure that Jack is the one who killed Mack.

4.3
The trap is hidden in the meaning of the word "shadow"—a shadow does not have color or degrees of darkness.

4.4
In this puzzle, one can be subtracted only once from twenty-five; after that it is subtracted from twenty-four, then from twenty-three, etc.

4.5
This puzzle, which simulates the style of an algebraic rate problem, can be solved correctly only if the meaning of the verb "meet" is considered. Obviously, when two trains meet, they are the same distance from New York. One train is going away from the city and the other toward it, but their distance from the city is the same.

4.6
The way in which this puzzle presents its facts induces solvers to conclude that 100 cats will kill 100 rats in 100 minutes. However, if the information of the puzzle is extracted from its deceptive style, the solution becomes obvious: if 3 cats kill 3 rats in 3 minutes, it must take each cat 3 minutes to kill a rat. So 100 cats will kill 100 rats—one kill each—in 3 minutes, and in general, n cats will kill n rats in 3 minutes. It does not matter how many cats are killing rats at the same time.

4.7
Similarly, in this puzzle, it will take 3¾ minutes to boil any number of eggs, assuming of course that the pot can hold them all.

4.8
In this puzzle, the key is to realize that each sister does not have a different brother. So the farmer had seven daughters and one son, who was the brother of each of the seven daughters. In total, he had eight children.

Chapter 5

5.1

Here is one possible magic square. It has been constructed by changing the order of the rows:

4	9	2
3	5	7
8	1	6

Figure S.16

Its mirror image is another:

2	9	4
7	5	3
6	1	8

Figure S.17

5.2

Let H stand for a husband and W for a wife, with subscript numbers indicating who is married to whom. The three couples and their crossings can be represented as follows:

Original Side	On the Boat	Other Side
0. H_1 W_1 H_2 W_2 H_3 W_3	— —	— — — — — —
1. — — H_2 W_2 H_3 W_3	H_1 W_1 →	— — — — — —
2. — — H_2 W_2 H_3 W_3	← W_1	H_1 — — — — —
3. — — H_2 — H_3 W_3	W_1 W_2 →	H_1 — — — — —
4. — — H_2 — H_3 W_3	← W_2	H_1 W_1 — — — —
5. — — — — H_3 W_3	H_2 W_2 →	H_1 W_1 — — — —
6. — — — — H_3 W_3	← W_2	H_1 W_1 H_2 — — —
7. — — — — H_3 —	W_2 W_3 →	H_1 W_1 H_2 — — —
8. — — — — H_3 —	← W_3	H_1 W_1 H_2 W_2 — —
9. — — — — — —	H_3 W_3 →	H_1 W_1 H_2 W_2 — —

5.3

Amount of rice to be weighed	Weight added to rice	Weight in other pan
1	—	1
2	1	3
3	—	3
4	—	3 + 1
5	3 + 1	9
6	3	9
7	3	9 + 1

etc.

5.4

First, we remove one of the seven balls and put it aside; then we put three balls on each pan—three on the left pan and three on the right. If the pans balance we have identified the culprit ball in only one weighing—the one put aside. However, we cannot assume this will happen. We must always assume the worst-case scenario. So let us suppose that one of the pans goes up. The three balls on that pan obviously include the culprit ball. So we discard the other three, as well as the one initially put aside. Next, we put one of the three suspect balls aside and place the other two on separate pans. Once again, if the pans balance, the culprit ball is the one on the side; if one of the pans goes up, it contains the culprit ball.

5.5

Here is Chuquet's solution, step by step:

1. Fill the 5-pint jar from the cask:

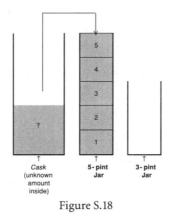

Figure S.18

2. Fill the 3-pint jar from the 5-pint one, leaving 2 pints in the 5-pint jar:

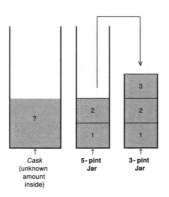

Figure S.19

3. Empty the 3-pint jar back into the cask:

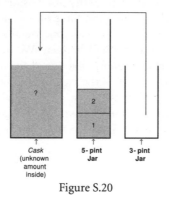

Figure S.20

4. Pour the 2 pints in the 5-pint jar into the 3-pint jar:

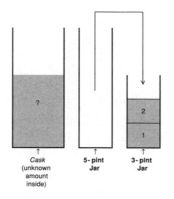

Figure S.21

5. Fill the 5-pint jar from the cask:

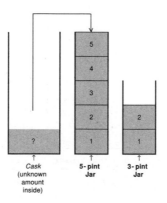

Figure S.22

6. Fill the 3-pint jar from the 5-pint jar. This will add a pint to the 3-pint jar and leave 4 pints in the 5-pint jar, which is the required solution.

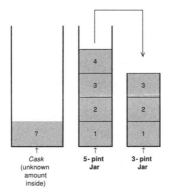

Figure S.23

5.6

Since Mack smoked only two-thirds of a cigarette, he would leave a butt equal to one-third of a cigarette. So, for every 3 cigarettes he smoked, he was able to piece together a new cigarette (⅓ butt + ⅓ butt + ⅓ butt = 1 new cigarette).

After smoking the original 27 cigarettes, he was thus able to make 9 new cigarettes. If we stop here, simply adding 27 (number of cigarettes Mack smoked originally) + 9 (number of new cigarettes made and smoked by Mack) = 36 (total number of cigarettes smoked by Mack), we will have ignored the fact that by smoking the 9 new cigarettes, Mack produced more butts. In fact, Mack's 9 new cigarettes produced 9 new butts of their own. From these 9 butts, Mack was able to make 3 more cigarettes (3 butts = 1 new cigarette). So, in addition to the 9 cigarettes Mack made from the original 27, he was also able to make 3 more from those 9. But those 3 extra cigarettes then produced 3 butts of their own, from which Mack was able to make one last cigarette. Altogether, therefore, Mack lit up 27 + 9 + 3 + 1 = 40 times before giving up his bad habit.

5.7

We start by labeling box BB with a 1, box WW with a 2, and box BW with a 3. A box will not contain the combination of ties that its label says it does, but it could contain any one of the other two combinations:

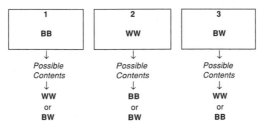

Figure S.24

Now, we start by drawing from box 1—the one incorrectly labeled BB. It may actually contain either BW or WW. If we draw a B, then we know that box 1 contains BW. If we draw a W, then we have learned nothing. We must, of course, assume the latter, worst-case scenario. But our second draw will tell us which of those two combinations it actually contains. We can then determine the actual contents of the other two boxes without drawing any ties from them.

Assume that we discovered that box 1 contained BW. Under the worst-case scenario, this would have taken 2 draws—one for W and the second one for B. Now there is no further need to draw from the other boxes because, by the process of elimination, box 2 must contain BB (not BW) and box 3, therefore, WW. Assume instead that box 1 contained WW. Again, under the worst-case scenario, this would have taken 2 draws—one for the first W and the second one for the other W. Once again, there is no need to draw from the other boxes because, by the process of elimination, box 3 must contain BB (not WW) and box 2, therefore, BW. So, if we draw from box 1, we will need at most two drawings to know what is in each box.

Now let us draw from box 2 instead—the one incorrectly labeled WW. It may actually contain either BB or BW. If we draw a W, then that box contains BW. If we draw a B, then we have learned nothing. As before, we must assume the latter, worst-case scenario. Our second draw will, of course, tell us which combination it actually contains. We can then deduce what the other boxes contain without drawing any ties from them. Let us assume that we drew BW from box 2. Now, by the process of elimination, box 1 must contain WW (not BW) and box 3, therefore, BB. If instead we drew BB, then by the process of elimination box 3 must contain WW (not BB) and box 1, therefore, BW. Once again, if we draw from box 2, we will need at most two drawings to know what is in each box.

Let us see what happens if we draw from box 3. It may contain either BB or WW. If we draw a B, then it contains BB. If we draw a W, then it contains WW. A second draw is therefore unnecessary, since we know that it will produce a match—either BB or WW. So in one drawing we will know what is in box 3. We can then deduce what the other boxes contain without drawing any ties from them. Assume that we drew B from box 3. It therefore contains BB. Now, by the process of elimination, box 2 must contain BW (not BB) and box 1, therefore, WW. If instead we drew W from box 3, then we know that it contains WW. By the process of elimination, box 1 must contain BW (not WW) and box 2, therefore, BB. In conclusion, we can determine the actual contents of each box in one draw if we make that draw from the one labeled BW.

5.8

Letting x represent Zerlina's age will lead nowhere, as readers can confirm for themselves. So we let x stand instead for the number of sons she has. Then the number of brothers each son has is $(x - 1)$. To grasp what this expression means in concrete terms, suppose that Zerlina has 5 sons—Andy, Bill, Charley, Dick, and Frank. How many brothers does Andy have? He has four brothers—Bill, Charley, Dick, and Frank. How many brothers does Bill have? He has four brothers—Andy, Charley, Dick, and Frank. And so on. Clearly, the number of brothers of each of Zerlina's five sons equals one less than the total number of her sons or, in algebraic terms, $(x-1)$.

We are told that each of Zerlina's sons has, himself, as many sons as he has brothers. Since each son has $(x-1)$ brothers, he also has $(x-1)$ sons. So Zerlina has x sons, and each one has $(x-1)$ sons of his own. This means that altogether she will have $x(x-1)$, or (x^2-x), grandsons. Finally, we are told that the combined number of Zerlina's sons (x) and grandsons (x^2-x) equals her age, which is a number between 50 and 70:

$x+(x^2-x)$ is a member of the set $\{50, 51, \ldots 70\}$

This simplifies to

x^2 is a member of the set $\{50, 51, \ldots 70\}$

The number we are looking for, therefore, is a square number between 50 and 70. The only such square number is $8^2 = 64$. So Zerlina is 64 years old.

5.9

We start by letting x represent the total number of sheep in the flock. Now, we can translate the relevant statements made by Betty into algebraic expressions as follows:

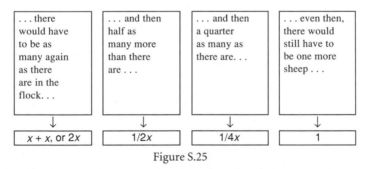

Figure S.25

If you are having difficulty seeing the reason behind the first expression—$x + x$ sheep—it is helpful to think about Betty's statement in concrete terms. Assume there are 25 sheep. How many sheep are "as many again"? Another 25, of course. So this is the number to be added to the original 25: i.e., $25 + 25 = 50$. The number 50 includes the original number of 25 and as many again as the original number: i.e., another 25.

When all these are added up, Betty states, the result is the 100 sheep that Bill originally thought were in the flock:

$2x + 1/2x + 1/4x + 1 = 100$

Solving for x, we get $x = 36$. So 36 is the number of sheep there were in the flock. This can be checked out by replacing Betty's verbal statements with actual figures:

The number there are in the flock	=	36
as many again	=	36
half as many more than there are	=	18 *(= half of 36)*
a quarter as many as there are	=	9 *(= one-quarter of 36)*
and one more sheep	=	+1
Total	=	100

Chapter 6

6.1

Removing the three rods labeled b, m, and o leaves the required three squares:

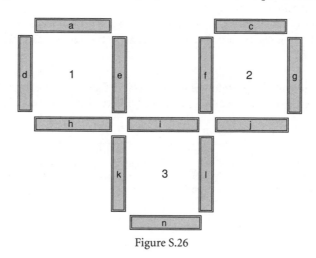

Figure S.26

6.2

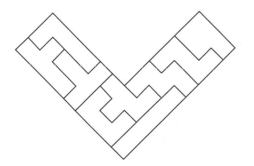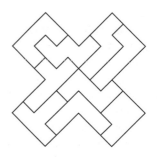

Figure S.27

6.3

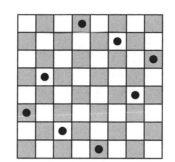

6.4

The probability of drawing an ace is $\frac{1}{13}$ and the probability of drawing a king is also $\frac{1}{13}$. The probability of drawing either one or the other is the sum of these two: $\frac{1}{13} + \frac{1}{13} = \frac{2}{13}$.

BIBLIOGRAPHY AND
GENERAL READING LIST

The following bibliography includes both the works cited in the text and, more generally, the works that have constituted the background to my treatment of the various topics.

Abbott, E. A. [1884] 1963. *Flatland: A Romance of Many Dimensions.* New York: Barnes and Noble.

Abraham, R. M. 1933. *Diversions and Pastimes: A Second Series of Winter Nights Entertainments.* London: Constable & Co. Reprint, New York: Dover, 1964.

Aczel, A. D. 2000. *The Mystery of the Aleph: Mathematics, the Kabbalah, and the Search for Infinity.* New York: Four Walls Eight Windows.

Agostini, F., and N. A. De Carlo. 1985. *Giochi della intelligenza.* Milan: Mondadori.

———. 1987. *Intelligence Games.* London: Macdonald.

Ahrens, W. 1921. *Mathematische Unterhalten und Spiele.* Leipzig: B. G. Teubner.

Andrews, W. S. 1960. *Magic Squares and Cubes.* New York: Dover.

Antenos-Conforti, E., E. Barbeau, and M. Danesi. 1997. *Problem-Solving in Mathematics: A Semiotic Perspective with Educational Implications.* Monograph Series of the Toronto Semiotic Circle, 20. Toronto: Toronto Semiotic Circle.

Aristotle. 1952. *Poetics.* In *The Works of Aristotle,* vol. 11, edited by W. D. Ross. Oxford: Clarendon Press.

Ascher, M. 1990. A River-Crossing Problem in Cross-Cultural Perspective. *Mathematics Magazine* 63: 26–29.

Atkinson, D. 1951. The Origin and Date of the "Sator" Word Square. *Journal of Ecclesiastical History* 2: 1–18.

Averbach, B., and O. Chein. 1980. *Problem Solving through Recreational Mathematics.* New York: Dover.

Bachet, C.-G. 1612. *Problèmes plaisans et délectables, qui se font par les nombres: Partie recueillis de divers autheurs, partie inventez de nouveau avec leur demonstration.* Lyon: chez P. Rigaud & associez.

Ball, W. W. Rouse. 1972. *Mathematical Recreations and Essays.* 12th ed., revised by H. S. M. Coxeter. Toronto: University of Toronto Press.

Barbeau, E. 1995. *After Math: Puzzles and Brainteasers.* Toronto: Wall and Emerson.

Barnsley, M. 1988. *Fractals Everywhere.* Boston: Academic.

Barwise, J., and J. Etchemendy. 1986. *The Liar.* Oxford: Oxford University Press.

Baudelaire, C. 1868. *Curiosités esthétiques.* Paris: Editions de la Pléiade.

Beasley, J. 1990. *The Mathematics of Games.* Oxford: Oxford University Press.

Beck, N. A. 1997. *Anti-Roman Cryptograms in the New Testament: Symbolic Messages of Hope and Liberation.* New York: Peter Lang.

Beckmann, P. 1971. *A History of Pi.* New York: St. Martin's.

Beiler, A. H. 1966. *Recreations in the Theory of Numbers: The Queen of Mathematics Entertains.* New York: Dover.

Bell, R. C., and M. Cornelius. 1988. *Board Games round the World: A Resource Book for Mathematical Investigations.* Cambridge: Cambridge University Press.

Benson, D. C. 1999. *The Moment of Proof: Mathematical Epiphanies.* Oxford: Oxford University Press.

Benson, W. H., and O. Jacoby. 1976. *New Recreations with Magic Squares.* New York: Dover.

———. 1981. *Magic Cubes: New Recreations.* New York: Dover.

Bergerson, H. W. 1973. *Palindromes and Anagrams.* New York: Dover.

Bergin, T. G., and M. Fisch. 1984. *The New Science of Giambattista Vico: Unabridged Translation of the Third Edition (1744).* Ithaca: Cornell University Press.

Berlekamp, E., and T. Rodgers, eds. 1999. *The Mathemagician and Pied Puzzler: A Collection in Tribute to Martin Gardner.* Natick, Mass.: A. K. Peters.

Blatner, D. 1997. *The Joy of Pi.* London: Penguin.

Block, J. R., and H. Yuker. 1992. *Can You Believe Your Eyes? Over 250 Illusions and Visual Oddities.* New York: Brunner/Mazel.

Bolt, B. 1982. *Mathematical Activities: A Resource Book for Teachers.* Cambridge: Cambridge University Press.

———. 1984. *The Amazing Mathematical Amusement Arcade.* Cambridge: Cambridge University Press.

———. 1987. *Even More Mathematical Activities.* Cambridge: Cambridge University Press.

Bombaugh, C. C. 1961. *Oddities and Curiosities of Words and Literature.* New York: Dover.

Boole, G. 1854. *An Investigation of the Laws of Thought.* Reprint, New York: Dover, 1958.

Bronowski, J. 1977. *A Sense of the Future: Essays in Natural Philosophy.* Cambridge, Mass.: MIT Press.

Brooke, M. 1963a. *150 Puzzles in Crypt-Arithmetic.* New York: Dover.

———. 1963b. *Coin Games and Puzzles.* New York: Dover.

Bunt, L., P. Jones, and J. Bedient. 1976. *The Historical Roots of Elementary Mathematics.* New York: Dover.

Butterworth, B. 1999. *The Mathematical Brain.* London: Macmillan.

Carroll, L. 1885. *A Tangled Tale.* London: Macmillan.

———. 1958a. *The Game of Logic.* New York: Dover.

———. 1958b. *Mathematical Recreations of Lewis Carroll.* New York: Dover.

———. 1958c. *Pillow Problems and a Tangled Tale.* New York: Dover.

———. 1990. *More Annotated Alice: Alice's Adventures in Wonderland and Through the Looking Glass.* With notes by M. Gardner. New York: Random House.

Cavalli-Sforza, L. L. 2000. *Genes, Peoples, and Languages.* New York: Farrar, Straus and Giroux.

Céard, J. 1986. *Rébus de la Renaissance: des images qui parlent.* Paris: Maisonneuve et Larose.

Chase, A. 1979. *The Rhind Mathematical Papyrus: Free Translation and Commentary with Selected Photographs, Transcriptions, Transliterations, and Literal Translations.* Reston, Va.: National Council of Teachers of Mathematics.

Chevalier, J., and A. Gheerbrant. 1994. *A Dictionary of Symbols.* Oxford: Blackwell.

Clawson, C. C. 1996. *Mathematical Mysteries: The Beauty and Magic of Numbers.* New York: Plenum Press.

Close, F. 2000. *Lucifer's Legacy: The Meaning of Asymmetry.* Oxford: Oxford University Press.

Cohen, M. N. 1995. *Lewis Carroll: A Biography.* New York: Alfred A. Knopf.

Costello, M. J. 1996. *The Greatest Puzzles of All Time.* New York: Dover.

Craze, R. 2001. *Numerology Decoder: Unlock the Power of Numbers to Reveal Your Innermost Desires and Foretell Your Future.* New York: Barron's Educational Series.

Dalgety, J., and E. Hordern. 1999. Classification of Mechanical Puzzles and Physical Objects Related to Puzzles. In *The Mathemagician and Pied Puzzler: A Collection in*

Tribute to Martin Gardner, edited by E. Berlekamp and T. Rodgers, pp. 175–186. Natick, Mass.: A. K. Peters.

Danesi, M. 1997. *Increase Your Puzzle IQ!* New York: John Wiley.

David, F. N. 1962. *Games, Gods, and Gambling: A History of Probability and Statistical Ideas.* New York: Dover.

Davis, P. J., and R. Hersh. 1986. *Descartes' Dream: The World According to Mathematics.* Boston: Houghton Mifflin.

Dawkins, R. 1976. *The Selfish Gene.* Oxford: Oxford University Press.

———. 1987. *The Blind Watchmaker.* New York: Norton.

———. 1995. *River out of Eden: A Darwinian View of Life.* New York: Basic Books.

———. 1998. *Unweaving the Rainbow: Science, Delusion, and the Appetite for Wonder.* Boston: Houghton Mifflin.

De Morgan, A. 1954. *A Budget of Paradoxes.* New York: Dover.

Dehaene, S. 1997. *The Number Sense: How the Mind Creates Mathematics.* Oxford: Oxford University Press.

Deregowski, J. B. 1972. Pictorial Perception and Culture. *Scientific American* 227: 82–88.

Devlin, K. 1997. *Mathematics, the Science of Patterns: The Search for Order in Life, Mind, and the Universe.* New York: Scientific American Library.

———. 1998a. *The Language of Mathematics: Making the Invisible Visible.* New York: W. H. Freeman.

———. 1998b. *Mathematics: The New Golden Age.* 2nd ed. London: Penguin.

———. 2000. *The Math Gene: How Mathematical Thinking Evolved and Why Numbers Are Like Gossip.* New York: Basic Books.

Doob, P. 1990. *The Idea of the Labyrinth from Classical Antiquity through the Middle Ages.* Ithaca: Cornell University Press.

Dörrie, H. 1965. *100 Great Problems in Elementary Mathematics.* New York: Dover.

Doxiadis, A. 2000. *Uncle Petros and Goldbach's Conjecture.* London: Faber and Faber.

Dudeney, H. E. 1926. *Modern Puzzles and How to Solve Them.* London: C. A. Pearson.

———. 1958a. *The Canterbury Puzzles and Other Curious Problems.* New York: Dover.

———. 1958b. *Amusements in Mathematics.* New York: Dover.

———. 1967. *536 Curious Problems and Puzzles.* New York: Scribner.

Eco, U. 1998. *Serendipities: Language and Lunacy.* Translated by William Weaver. New York: Columbia University Press.

Eiss, H. E. 1988. *Dictionary of Mathematical Games, Puzzles, and Amusements.* New York: Greenwood.

Falkener, E. 1892. *Games Ancient and Oriental and How to Play Them: Being the Games of the Ancient Egyptians, the Hiera Gramme of the Greeks, the Ludus Latrunculorum of the Romans, and the Oriental Games of Chess, Draughts, Backgammon and Magic Squares.* London: Longmans, Green.

Falletta, N. 1983. *The Paradoxicon: A Collection of Contradictory Challenges, Problematic Puzzles, and Impossible Illustrations.* New York: John Wiley.

Fatsis, S. 2001. *Word Freak: Heartbreak, Triumph, Genius, and Obsession in the World of Competitive Scrabble Players.* Boston: Houghton Mifflin.

Fauvel, J., R. Flood, and R. Wilson, eds. 1993. *Möbius and His Band: Mathematics and Astronomy in Nineteenth-Century Germany.* Oxford: Oxford University Press.

Fisher, A., and G. Gerster. 1990. *The Art of the Maze.* London: George Weidenfeld and Nicolson.

Frey, A. H., and D. Singmaster. 1982. *Handbook of Cubik Math.* Hillside, N.J.: Enslow.

Gaines, H. F. 1956. *Cryptanalysis: A Study of Ciphers and Their Solution.* New York: Dover.

Galilei, G. 1914. *Dialogues concerning Two New Sciences.* New York: Macmillan.

Gardiner, A. 1987. *Mathematical Puzzling.* New York: Dover.

Gardner, M. 1956. *Mathematics, Magic, and Mystery.* New York: Dover.

———. 1959. *The Scientific American Book of Mathematical Puzzles and Diversions.* New York: Simon and Schuster.

———. 1961a. *The Second Scientific American Book of Mathematical Puzzles.* New York: Simon and Schuster.

———. 1961b. *Entertaining Mathematical Puzzles.* New York: Dover.

———. 1966. *More Mathematical Puzzles and Diversions.* Harmondsworth: Penguin.

———. 1969. *The Unexpected Hanging, and Other Mathematical Diversions.* New York: Simon and Schuster.

———. 1971. *Martin Gardner's Sixth Book of Mathematical Games from Scientific American.* San Francisco: Freeman.

———. 1972. *Codes, Ciphers, and Secret Writing.* New York: Dover.

———. 1979a. *Aha! Insight!* New York: Scientific American.

———. 1979b. *Mathematical Circus.* New York: Knopf.

———. 1982. *Aha! Gotcha! Paradoxes to Puzzle and Delight.* San Francisco: Freeman.

———. 1983. *Wheels, Life, and Other Mathematical Amusements.* New York: Freeman.

———. 1986. *Knotted Doughnuts and Other Mathematical Entertainments.* New York: Freeman.

———. 1987. *Riddles of the Sphinx and Other Mathematical Puzzle Tales.* Washington, D.C.: Mathematical Association of America.

———. 1988. *Time Travel and Other Bewilderments.* New York: Freeman.

———. 1989a. *Mathematical Carnival.* Washington, D.C.: Mathematical Association of America.

———. 1989b. *Penrose Tiles to Trapdoor Ciphers.* New York: Freeman.

———. 1990. *Mathematical Magic Show.* Washington, D.C.: Mathematical Association of America.

———. 1992. *Fractal Music, Hypercards, and More.* New York: Freeman.

———. 1994a. *My Best Mathematical and Logic Puzzles.* New York: Dover.

———. 1994b. *New Mathematical Diversions.* Washington, D.C.: Mathematical Association of America.

———. 1996. *The Universe in a Handkerchief: Lewis Carroll's Mathematical Recreations, Games, Puzzles, and Word Plays.* New York: Copernicus.

———. 1997. *The Last Recreations: Hydras, Eggs, and Other Mathematical Mystifications.* New York: Copernicus.

———. 1998. A Quarter-Century of Recreational Mathematics. *Scientific American* 279: 68–75.

———. 2001. *The Colossal Book of Mathematics.* New York: W. W. Norton.

Gillam, B. 1980. Geometrical Illusions. *Scientific American* 242: 102–111.

Gillings, R. J. 1961. Think-of-a-Number: Problems 28 and 29 of the Rhind Mathematical Papyrus. *The Mathematics Teacher* 54: 97–102.

———. 1962. Problems 1 to 6 of the Rhind Mathematical Papyrus. *The Mathematics Teacher* 55: 61–65.

———. 1972. *Mathematics in the Time of the Pharaohs.* Cambridge, Mass.: MIT Press.

Gleick, J. 1987. *Chaos: Making a New Science.* New York: Viking.

Glenn, W. H., and D. A. Johnson. 1961. *Invitation to Mathematics.* New York: Dover.

Gödel, K. 1931. Über formal unentscheidbare Sätze der Principia Mathematica und verwandter Systeme, Teil I. *Monatshefte für Mathematik und Physik* 38: 173–189.

Golomb, S. W. 1965. *Polyominoes.* New York: Scribner.

———. 1996. *Polyominoes: Puzzles, Patterns, Problems.* Princeton: Princeton University Press.

Gregory, R. 1997. *Mirrors in Mind.* New York: W. H. Freeman.

Grimal, P. 1963. *Dictionnaire de la mythologie grecque et romaine*. 3rd ed. Paris: Presses Universitaires de France.

Gullberg, J. 1997. *Mathematics from the Birth of Numbers*. New York: W. W. Norton.

Haken, W. 1977. Every Planar Map Is Four-Colorable. *Illinois Journal of Mathematics* 21: 429–567.

Haken, W., and K. Appel. 1977. The Solution of the Four-Color-Map Problem. *Scientific American* 237: 108–121.

Hall, M. P. 1973. *The Secret Teachings of All Ages*. Los Angeles: Philosophical Research Society.

Hannas, L. 1972. *The English Jigsaw Puzzle, 1760–1890*. London: Wayland.

———. 1981. *The Jigsaw Book*. New York: Dial.

Harrowitz, N. 1983. The Body of the Detective Model: Charles S. Peirce and Edgar Allan Poe. In *Dupin, Holmes, Peirce: The Sign of Three*, edited by U. Eco and T. A. Sebeok, pp. 189–195. Bloomington: Indiana University Press.

Heath, R. V. 1953. *Mathemagic: Magic, Puzzles, and Games with Numbers*. New York: Dover.

Heath, T. L. 1958. *The Works of Archimedes with the Method of Archimedes*. New York: Dover.

Heesterman, J. C. 1997. On Riddles. In *Significations: Essays in Honour of Henry Schogt*, edited by P. Bhatt, pp. 65–69. Toronto: Canadian Scholars' Press.

Ho Peng Yoke. 1985. *Li, Qi, and Shu: An Introduction to Science and Civilization in China*. Mineola, N.Y.: Dover.

Hoffman, D. D. 1983. The Interpretation of Visual Illusions. *Scientific American* 249: 154–162.

———. 1998. *Visual Intelligence: How We Create What We See*. New York: W. W. Norton.

Hofstadter, D. 1979. *Gödel, Escher, Bach: An Eternal Golden Braid*. New York: Basic Books.

Holroyd, S., and N. Powell. 1991. *Mysteries of Magic*. London: Bloomsbury Books.

Holt, M. 1978. *Math Puzzles and Games*. New York: Walker.

Holt, M. J., and A. J. McIntosh. 1966. *The Scope of Mathematics: A Fresh Look at Mathematics for the Non-specialist*. Oxford: Oxford University Press.

Honsberger, R. 1973. *Mathematical Gems*. Washington, D.C.: Mathematical Association of America.

———. 1978. *Mathematical Morsels*. Washington, D.C.: Mathematical Association of America.

Hooke, S. H. 1935. *The Labyrinth: Further Studies in the Relation between Myth and Ritual in the Ancient World*. London: Society for Promoting Christian Knowledge.

Hovanec, H. 1978. *The Puzzler's Paradise: From the Garden of Eden to the Computer Age*. New York: Paddington.

Hudson, D. 1954. *Lewis Carroll: An Illustrated Biography*. London: Constable.

Huizinga, J. 1924. *The Waning of the Middle Ages: A Study of the Forms of Life, Thought and Art in France and the Netherlands in the XIVth and XVth Centuries*. Garden City, N.J.: Doubleday.

Hunter, J. A. H. 1965. *Fun with Figures*. New York: Dover.

Hunter, J. A. H., and J. S. Madachy. 1963. *Mathematical Diversions*. New York: Van Nostrand.

Huntley, H. E. 1970. *The Divine Proportion: A Study in Mathematical Beauty*. New York: Dover.

James. P. D. 1986. Interview. *Face*, 21–24.

Joseph, G. G. 1991. *The Crest of the Peacock: Non-European Roots of Mathematics*. London: Penguin.

Jourdain, P. E. B. 1913. Tales with Philosophical Morals. *Open Court* 27: 310–315.

Kahn, D. 1967. *The Codebreakers: The Story of Secret Writing*. New York: Macmillan.

Kaplan, R. 1998. *The Nothing That Is: A Natural History of Zero*. Oxford: Oxford University Press.

Kappraff, J. 1991. *Connections: The Geometric Bridge between Art and Science*. New York: McGraw-Hill.

Kasner, E., and J. Newman. 1940. *Mathematics and the Imagination*. New York: Simon and Schuster.

Kirkman, T. P. 1847. A Schoolgirl Problem. *Cambridge and Dublin Mathematics Journal* 2: 191–204.

Klarner, D. A., ed. 1981. *Mathematical Recreations: A Collection in Honor of Martin Gardner*. Mineola, N.Y.: Dover.

Kline, M. 1959. *Mathematics and the Physical World*. New York: Crowell.

———. 1985. *Mathematics and the Search for Knowledge*. Oxford: Oxford University Press.

Knuth, D. E. 1999. Biblical Ladders. In *The Mathemagician and Pied Puzzler: A Collection in Tribute to Martin Gardner*, edited by E. Berlekamp and T. Rodgers, pp. 29–34. Natick, Mass.: A. K. Peters.

Koch, S. 1981. The Nature and Limits of Psychological Knowledge: Lessons of a Century Qua "Science." *American Psychologist* 36: 260–265.

Kordemsky, B. A. 1972. *The Moscow Puzzles: 359 Mathematical Recreations*. Harmondsworth: Penguin.

Kraitchik, M. 1953. *Mathematical Recreations*. New York: Dover.

Lakoff, G., and R. E. Núñez. 2000. *Where Mathematics Comes From: How the Embodied Mind Brings Mathematics into Being*. New York: Basic Books.

Langer, S. 1948. *Philosophy in a New Key: A Study in the Symbolism of Reason, Rite and Art*. New York: Penguin Books.

Lawler, J. M. 1978. *IQ, Heritability, and Racism*. New York: International Publishers.

Lemon, D. 1890. *Everybody's Illustrated Book of Puzzles*. London: Saxon.

Lewis, A. J. 1893. *Puzzles Old and New*. London: Frederick Warne.

Li Yan and Du Shiran. 1987. *Chinese Mathematics: A Concise History*. Translated by J. H. Crossley and A. W.-C. Lun. Oxford: Clarendon Press.

Lindgren, H., and G. Frederickson. 1972. *Recreational Problems in Geometric Dissections and How to Solve Them*. New York: Dover.

Lobosco, M. L. 1998. *Mental Math Workout*. New York: Sterling.

Lockridge, R. 1941. *The Labyrinth*. Reprint, Westport: Hyperion, 1975.

Loyd, S. 1914. *Cyclopedia of Puzzles*. New York: Lamb.

———. 1959–1960. *Mathematical Puzzles of Sam Loyd*. 2 vols., compiled by M. Gardner. New York: Dover.

———. [1903] 1968. *The Eighth Book of Tan*. New York: Dover.

Lucas, F. E. A. 1882–1894. *Récreations mathématiques*. 4 vols. Paris: Gauthier-Villars.

Luckiesh, M. 1965. *Visual Illusions: Their Causes, Characteristics, and Applications*. New York: Dover.

Lundy, M. 1998. *Sacred Geometry*. New York: Walker.

Madachy, J. S. 1966. *Mathematics on Vacation*. New York: Scribner.

Mahavira. 1912. *The Ganita-Sara-Sangraha of Mahaviracarya*. Translated by M. Rangacarya. Madras: Government Press.

Mandelbrot, B. 1977. *The Fractal Geometry of Nature*. San Francisco: Freeman.

Maor, E. 1987. *To Infinity and Beyond: A Cultural History of the Infinite*. Boston: Birkhäuser.

———. 1998. *Trigonometric Delights*. Princeton: Princeton University Press.

Mathews, W. H. 1970. *Mazes and Labyrinths: Their History and Development*. New York: Dover.

Meehan, A. 1993. *Maze Patterns*. New York: Thames and Hudson.

Millington, R. 1974. *The Strange World of the Crossword.* London: M. & J. Hobbs. Published in the U.S. as *Crossword Puzzles: Their History and Their Cult.* New York: Thomas Nelson, 1974.

Mishlove, J. 1993. *The Roots of Consciousness.* New York: Marlowe.

Mohr, M. S. 1993. *The Games Treasury: More Than 300 Indoor and Outdoor Favorites with Strategies, Rules, and Traditions.* Shelburne, Vt.: Chapters.

Morris, D. 1967. *The Naked Ape.* London: Cape.

———. 1969. *The Human Zoo.* London: Cape.

Moscovich, I. 2001. *1000 PlayThinks.* New York: Workman Publishing Co.

Mott-Smith, G. 1978. *Mathematical Puzzles.* New York: Dover.

Neugebauer, O. 1957. *The Exact Sciences in Antiquity.* New York: Dover.

Neugebauer, O., A. Sachs, and Albrecht Götze. 1945. *Mathematical Cuneiform Texts.* New Haven: American Oriental Society and American Schools of Oriental Research.

Neuwirth, L. 1979. The Theory of Knots. *Scientific American* 240: 110–124.

Northrop, E. P. 1944. *Riddles in Mathematics: A Book of Paradoxes.* Harmondsworth: Penguin.

O'Beirne, T. H. 1965. *Puzzles and Paradoxes.* Oxford: Oxford University Press.

Ogilvy, C. S. 1956. *Excursions in Mathematics.* New York: Dover.

Olivastro, D. 1993. *Ancient Puzzles: Classic Brainteasers and Other Timeless Mathematical Games of the Last 10 Centuries.* New York: Bantam.

Pappas, T. 1989. *The Joy of Mathematics: Discovering Mathematics All around You.* San Carlos: Wide World Publishing.

———. 1991. *More Joy of Mathematics: Exploring Mathematics All around You.* San Carlos: Wide World Publishing.

———. 1993. *Fractals, Googols, and Other Mathematical Tales.* San Carlos: Wide World Publishing.

———. 1999. *Mathematical Footprints: Discovering Mathematical Impressions All around Us.* San Carlos: Wide World Publishing.

Paulos, J. A. 1980. *Mathematics and Humor.* Chicago: University of Chicago Press.

———. 1985. *I Think, Therefore I Laugh: An Alternative Approach to Philosophy.* New York: Columbia University Press.

———. 1991. *Beyond Innumeracy.* New York: Vintage.

Peace, N. 1991. *Mathematical Games and Puzzles.* Thornhill: Tynron.

Pedoe, D. 1976. *Geometry and the Visual Arts.* New York: Dover.

Peet, T. E. 1923. *The Rhind Mathematical Papyrus.* Liverpool: University Press of Liverpool.

Peirce, C. S. 1923. *Chance, Love, and Logic: Philosophical Essays.* New York: Harcourt, Brace.

Penrose, L. S., and R. Penrose. 1958. Impossible Objects: A Special Type of Visual Illusion. *British Journal of Psychology* 49: 31–33.

Perelman, Y. I. 1979. *Algebra Can Be Fun.* Moscow: MIR.

Perkins, D. 1995. *Outsmarting IQ: The Emerging Science of Learnable Intelligence.* New York: Free Press.

———. 2000. *Archimedes' Bathtub: The Art and Logic of Breakthrough Thinking.* New York: W. W. Norton.

Peterson, I., and N. Henderson. 2000. *Math Trek: Adventures in the Math Zone.* New York: John Wiley.

Phillips, H. 1933. *The Playtime Omnibus.* London: Faber and Faber.

———. 1934. *The Sphinx Problem Book.* London: Faber and Faber.

———. 1936. *Brush Up Your Wits.* London: Dent.

———. 1937. *Question Time: An Omnibus of Problems for a Brainy Day.* London: Dent.

———. 1966. *Caliban's Problem Book: Mathematical, Inferential and Crytographic Puzzles.* New York: Dover.

Pickover, C. A. 2002. *The Zen of Magic Squares, Circles, and Stars.* Princeton: Princeton University Press.

Pinker, S. 1997. *How the Mind Works.* New York: Norton.

Poe, E. A. 1960. *The Fall of the House of Usher and Other Tales.* New York: Signet Classics.

Pólya, G. 1957. *How to Solve It: A New Aspect of Mathematical Method.* New York: Doubleday.

Rademacher, H. 1957. *The Enjoyment of Mathematics: Selections from Mathematics for the Amateur.* Princeton: Princeton University Press.

Ray-Chaudhuri, D. K., and R. M. Wilson. 1971. Solution of Kirkman's Schoolgirl Problem. *Combinatorics* 19: 187–203.

Read, R. C. 1965. *Tangrams: 330 Puzzles.* New York: Dover.

Rescher, N. 2001. *Paradoxes: Their Roots, Range and Resolution.* Chicago and La Salle: Open Court.

Richards, D. 1999. Martin Gardner: A "Documentary." In *The Mathemagician and Pied Puzzler: A Collection in Tribute to Martin Gardner,* edited by E. Berlekamp and T. Rodgers, pp. 3–12. Natick, Mass.: A. K. Peters.

Richards, I. A. 1936. *The Philosophy of Rhetoric.* Oxford: Oxford University Press.

Rodgers, N. 1998. *Incredible Optical Illusions: A Spectacular Journey through the World of the Impossible.* London: Quarto.

Rosenheim, S. J. 1997. *The Cryptographic Imagination: Secret Writing from Edgar Poe to the Internet.* Baltimore: Johns Hopkins University Press.

Rósza, P. 1957. *Playing with Infinity: Mathematical Explorations and Excursions.* New York: Dover.

Rubik, E. 1987. *Rubik's Cube Compendium.* Oxford: Oxford University Press.

Rucker, R. 1987. *Mind Tools: The Five Levels of Mathematical Reality.* Boston: Houghton Mifflin.

Salem, L., F. Testard, and C. Salem. 1992. *The Most Beautiful Mathematical Formulas.* New York: John Wiley.

Salmon, W. C., ed. 1970. *Zeno's Paradoxes.* Indianapolis: Bobbs-Merrill.

Salny, A. F., and L. B. Frumkes. 1986. *Mensa Think-Smart Book.* New York: Harper and Row.

Sampson, G. 1997. *Educating Eve: The "Language Instinct" Debate.* London: Cassell.

Sapir, E. 1921. *Language.* New York: Harcourt, Brace, and World.

Sardar, Z., J. Ravetz, and B. Van Loon. 1999. *Introducing Mathematics.* Cambridge: Icon Books.

Sawyer, W. W. 1955. *Prelude to Mathematics.* Harmondsworth: Penguin.

Schattschneider, D. 1990. *Visions of Symmetry: Notebooks, Periodic Drawings, and Related Work of M.C. Escher.* New York: Freeman.

Schimmel, R. 1993. *The Mystery of Numbers.* Oxford: Oxford University Press.

Schuh, F. 1968. *The Master Book of Mathematical Recreations.* New York: Dover.

Scott, Charles T. 1965. *Persian and Arabic Riddles: A Language-Centered Approach to Genre Definition.* The Hague: Mouton.

Searle, J. R. 1999. *Mind, Language, Society: Philosophy in the Real World.* New York: Basic Books.

Sebeok, T. A., and J. Umiker-Sebeok. 1983. You Know My Method. In *Dupin, Holmes, Peirce: The Sign of Three,* edited by Umberto Eco and Thomas A. Sebeok, pp. 3–42. Bloomington: Indiana University Press.

Segerstråle, U. 2000. *Defenders of the Truth: The Battle for Science in the Sociobiology Debate and Beyond.* Oxford: Oxford University Press.

Seife, C. 2000. *Zero: The Biography of a Dangerous Idea.* Harmondsworth: Penguin.

Shepard, R. N. 1990. *Mind Sights: Original Visual Illusions, Ambiguities, and Other Anomalies.* New York: W. H. Freeman.

Simon, S. 1984. *The Optical Illusion Book.* New York: William Morrow.

Simon, W. 1964. *Mathematical Magic.* New York: Dover.

Singh, S. 1997. *Fermat's Enigma: The Quest to Solve the World's Greatest Mathematical Problem.* New York: Walker.

———. 1999. *The Code Book: The Evolution of Secrecy from Mary, Queen of Scots to Quantum Cryptography.* New York: Doubleday.

Singmaster, D. 1999. Some Diophantine Recreations. In *The Mathemagician and Pied Puzzler: A Collection in Tribute to Martin Gardner,* edited by E. Berlekamp and T. Rodgers, pp. 219–235. Natick, Mass.: A. K. Peters.

Slocum, J., and J. Botermans. 1992. *New Book of Puzzles: 101 Classic and Modern Puzzles to Make and Solve.* New York: Freeman.

———. 1994. *The Book of Ingenious and Diabolical Puzzles.* New York: Times Books.

Smullyan, R. 1978. *What Is the Name of This Book? The Riddle of Dracula and Other Logical Puzzles.* Englewood Cliffs, N.J.: Prentice-Hall.

———. 1979. *The Chess Mysteries of Sherlock Holmes.* New York: Knopf.

———. 1982a. *Alice in Puzzle-Land.* Harmondsworth: Penguin.

———. 1982b. *The Lady or the Tiger? and Other Logic Puzzles.* New York: Knopf.

———. 1985. *To Mock a Mocking Bird and Other Logic Puzzles.* New York: Knopf.

———. 1987. *Forever Undecided: A Puzzle Guide to Gödel.* New York: Knopf.

———. 1997. *The Riddle of Scheherazade and Other Amazing Puzzles, Ancient and Modern.* New York: Knopf.

Sternberg, R. J. 1985. *Beyond IQ: A Triarchic Theory of Human Intelligence.* New York: Cambridge University Press.

Sternberg, R. J., and J. E. Davidson. 1982. The Mind of the Puzzler. *Psychology Today,* June 1982, pp. 37–44.

Stewart, I. 1987. *From Here to Infinity: A Guide to Today's Mathematics.* Oxford: Oxford University Press.

———. 1989. *Game, Set, and Math: Enigmas and Conundrums.* Cambridge: Basil Blackwell.

———. 1997. *The Magical Maze: Seeing the World through Mathematical Eyes.* New York: John Wiley.

———. 2001. *Flatterland: Like Flatland, only More So.* New York: Perseus Publishing.

Stewart, I., and J. Cohen. 1997. *Figments of Reality: The Evolution of the Curious Mind.* Cambridge: Cambridge University Press.

Swetz, F. J., and T. I. Kao. 1977. *Was Pythagoras Chinese? An Examination of Right Triangle Theory in Ancient China.* University Park: Pennsylvania State University Press.

Takagi, S. 1999. Japanese Tangram: The Sei Shonagon Pieces. In *The Mathemagician and Pied Puzzler: A Collection in Tribute to Martin Gardner,* edited by E. Berlekamp and T. Rodgers, pp. 97–98. Natick, Mass.: A. K. Peters.

Taylor, A. 1948. *The Literary Riddle before 1600.* Berkeley: University of California Press.

———. 1951. *English Riddles from Oral Tradition.* Berkeley: University of California Press.

Townsend, C. B. 1986. *The World's Best Puzzles.* New York: Sterling.

Trigg, C. W. 1985. *Mathematical Quickies: 270 Stimulating Problems with Solutions.* New York: Dover.

———. 1987. What Is Recreational Mathematics? *Mathematics Magazine* 51: 18–21.

Turing, A. 1936. On Computable Numbers with an Application to the Entscheidungs Problem. *Proceedings of the London Mathematical Society* 41: 230–265.

———. 1963. Computing Machinery and Intelligence. In *Computers and Thought,* edited by E. A. Feigenbaum and J. Feldman, pp. 123–134. New York: McGraw-Hill.

Tymoczko, T. 1979. The Four-Color Problem and Its Philosophical Significance. *Journal of Philosophy* 24: 57–83.

Van Delft, P., and J. Botermans. 1995. *Creative Puzzles of the World.* Berkeley: Key Curriculum Press.

Vernadore, J. 1991. Pascal's Triangle and Fibonacci Numbers. *The Mathematics Teacher* 84: 314–316.

Vorderman, C. 1996. *How Math Works.* Pleasantville: Reader's Digest Association.

Wakeling, E., ed. 1992. *Lewis Carroll's Games and Puzzles.* New York: Dover.

Wells, D. 1988. *Hidden Connections, Double Meanings: A Mathematical Exploration.* Cambridge: Cambridge University Press.

———. 1992. *The Penguin Book of Curious and Interesting Puzzles.* Harmondsworth: Penguin.

———. 1995. *You Are a Mathematician.* Harmondsworth: Penguin.

White, A. C. 1913. *Sam Loyd and His Chess Problems.* Leeds: Whitehead and Miller.

Whorf, B. L. 1956. *Language, Thought, and Reality: Selected Writings.* Edited by J. B. Carroll. Cambridge, Mass.: MIT Press.

Wiener, N. 1949. *Cybernetics, or Control and Communication in the Animal and the Machine.* Cambridge, Mass.: MIT Press.

Wilde, O. 1891. *The Picture of Dorian Gray.* London: Penguin.

Wilson, E. O., and M. Harris. 1981. Heredity versus Culture: A Debate. In *Anthropological Realities: Readings in the Science of Culture,* edited by J. Guillemin, pp. 450–465. New Brunswick, N.J.: Transaction Books.

Wright, R. 1994. *The Moral Animal: Evolutionary Psychology and Everyday Life.* New York: Pantheon.

Zebrowski, E. 1999. *A History of the Circle: Mathematical Reasoning and the Physical Universe.* New Brunswick: Rutgers University Press.

Zeki, S. 1999. *Inner Vision: An Exploration of Art and the Brain.* Oxford: Oxford University Press.

INDEX

Marcel Danesi is Professor of Semiotics and Anthropology at the University of Toronto and Director of the Program in Semiotics and Communication Theory. His publications include *Vico, Metaphor, and the Origin of Language* (Indiana University Press, 1993); *Cool: The Signs and Meanings of Adolescence* (1994); *Increase Your Puzzle IQ* (1997); *Analyzing Cultures* (Indiana University Press, 1999); *Of Cigarettes, High Heels, and Other Interesting Things: An Introduction to Semiotics* (1999); *The Forms of Meaning: Modeling Systems Theory and Semiotic Analysis* (with Thomas A. Sebeok, 2000); and *Encyclopedic Dictionary of Semiotics, Media, and Communications* (2000).